Cities and Low Carbon Transitions

Current societies face unprecedented risks and challenges connected to climate change. Addressing them will require fundamental transformations in the infrastructures that sustain everyday life, such as energy, water, waste and mobility. A transition to a 'low carbon' future implies a large scale reorganisation in the way societies produce and use energy. Cities are critical in this transition because they concentrate social and economic activities that produce climate change related emissions. At the same time, cities are increasingly recognised as sources of opportunities for climate change mitigation. Whether, how and why low carbon transitions in urban systems take place in response to climate change will therefore be decisive for the success of global mitigation efforts. As a result, climate change increasingly features as a critical issue in the management of urban infrastructure and in urbanisation policies.

Cities and Low Carbon Transitions presents a ground-breaking analysis of the role of cities in low carbon socio-technical transitions. Insights from the fields of urban studies and technological transitions are combined to examine how, why and with what implications cities bring about low carbon transitions. The book outlines the key concepts underpinning theories of socio-technical transition and assesses its potential strengths and limits for understanding the social and technological responses to climate change that are emerging in cities. It draws on a diverse range of examples including world cities, ordinary cities and transition towns, from North America, Europe, South Africa and China, to provide evidence that expectations, aspirations and plans to undertake purposive socio-technical transitions are emerging in different urban contexts.

This collection adds to existing literature on cities and energy transitions and introduces critical questions about power and social interests, lock-in and development trajectories, social equity and economic development, and socio-technical change in cities. The book addresses academics, policy makers, practitioners and researchers interested in the development of systemic responses in cities to curb climate change.

Harriet Bulkeley is a Professor at the Department of Geography, and Deputy Director of Durham Energy Institute, Durham University. Her research interests focus on the nature and politics of environmental governance and focuses on climate change, energy and urban sustainability.

Vanesa Castán Broto is a Lecturer at the Faculty of the Built Environment, University College London. Her research interests focus on how technology and environmental knowledge mediate the relationship between society and the environment.

Mike Hodson is Associate Director and Senior Research Fellow at the SURF Centre, University of Salford. His research interests focus on urban and regional transitions to low-carbon economies and understanding the lessons to be learned from such processes.

Simon Marvin is the Carillion Chair of Low Carbon Cities, Professor at the Department of Geography, Durham University and Deputy Director of Durham Energy Institute. His research interests focus on the changing relations between cities and infrastructure networks.

Routledge Studies in Human Geography

This series provides a forum for innovative, vibrant, and critical debate within Human Geography. Titles will reflect the wealth of research which is taking place in this diverse and ever-expanding field.

Contributions will be drawn from the main sub-disciplines and from innovative areas of work which have no particular sub-disciplinary allegiances.

Published:

1. **A Geography of Islands**
   Small island insularity
   Stephen A. Royle

2. **Citizenships, Contingency and the Countryside**
   Rights, culture, land and the environment
   Gavin Parker

3. **The Differentiated Countryside**
   Jonathan Murdoch, Philip Lowe, Neil Ward and Terry Marsden

4. **The Human Geography of East Central Europe**
   David Turnock

5. **Imagined Regional Communities**
   Integration and sovereignty in the global south
   James D Sidaway

6. **Mapping Modernities**
   Geographies of Central and Eastern Europe 1920–2000
   Alan Dingsdale

7. **Rural Poverty**
   Marginalisation and exclusion in Britain and the United States
   Paul Milbourne

8. **Poverty and the Third Way**
   Colin C. Williams and Jan Windebank

9. **Ageing and Place**
   Edited by Gavin J. Andrews and David R. Phillips

10. **Geographies of Commodity Chains**
    Edited by Alex Hughes and Suzanne Reimer

11. **Queering Tourism**
Paradoxical performances at Gay Pride parades
Lynda T. Johnston

12. **Cross-Continental Food Chains**
Edited by Niels Fold and Bill Pritchard

13. **Private Cities**
Edited by Georg Glasze, Chris Webster and Klaus Frantz

14. **Global Geographies of Post Socialist Transition**
Tassilo Herrschel

15. **Urban Development in Post-Reform China**
Fulong Wu, Jiang Xu and Anthony Gar-On Yeh

16. **Rural Governance**
International perspectives
Edited by Lynda Cheshire, Vaughan Higgins and Geoffrey Lawrence

17. **Global Perspectives on Rural Childhood and Youth**
Young rural lives
Edited by Ruth Panelli, Samantha Punch, and Elsbeth Robson

18. **World City Syndrome**
Neoliberalism and inequality in Cape Town
David A. McDonald

19. **Exploring Post Development**
Aram Ziai

20. **Family Farms**
Harold Brookfield and Helen Parsons

21. **China on the Move**
Migration, the state, and the household
C. Cindy Fan

22. **Participatory Action Research Approaches and Methods**
Connecting people, participation and place
Edited by Sara Kindon, Rachel Pain and Mike Kesby

23. **Time-Space Compression**
Historical geographies
Barney Warf

24. **Sensing Cities**
Monica Degen

25. **International Migration and Knowledge**
Allan Williams and Vladimir Baláž

26. **The Spatial Turn**
Interdisciplinary perspectives
Edited by Barney Warf and Santa Arias

27. **Whose Urban Renaissance?**
An international comparison of urban regeneration policies
Edited by Libby Porter and Katie Shaw

28. **Rethinking Maps**
Edited by Martin Dodge, Rob Kitchin and Chris Perkins

29. **Rural–Urban Dynamics**
Livelihoods, mobility and markets in African and Asian Frontiers
Edited by Jytte Agergaard, Niels Fold and Katherine V. Gough

30. **Spaces of Vernacular Creativity**
    Rethinking the cultural economy
    Edited by Tim Edensor, Deborah Leslie, Steve Millington and Norma Rantisi

31. **Critical Reflections on Regional Competitiveness**
    Gillian Bristow

32. **Governance and Planning of Mega-City Regions**
    An international comparative perspective
    Edited by Jiang Xu and Anthony G.O. Yeh

33. **Design Economies and the Changing World Economy**
    Innovation, production and competitiveness
    John Bryson and Grete Rustin

34. **Globalization of Advertising**
    Agencies, cities and spaces of creativity
    James Faulconbridge, Peter J. Taylor, J.V. Beaverstock and C Nativel

35. **Cities and Low Carbon Transitions**
    Edited by Harriet Bulkeley, Vanesa Castán Broto, Mike Hodson and Simon Marvin

36. **Globalization, Modernity and The City**
    John Rennie Short

37. **Climate Change and the Crisis of Capitalism**
    A chance to reclaim, self, society and nature
    Edited by Mark Pelling, David Manual Navarette and Michael Redclift

38. **New Economic Spaces in Asian Cities**
    From industrial restructuring to the cultural turn
    Edited by Peter W. Daniels, Kong Chong Ho and Thomas A. Hutton

39. **Cities, Regions and Flows**
    Edited by Peter V. Hall and Markus Hesse

40. **The Politics of Urban Cultural Policy**
    Global perspectives
    Edited by Carl Grodach and Daniel Silver

41. **Ecologies and Politics of Health**
    Edited by Brian King and Kelley Crews

42. **Producer Services in China**
    Economic and urban development
    Edited by Anthony G.O. Yeh and Fiona F. Yang

43. **Locating Right to the City in the Global South**
    Tony Roshan Samara, Shenjing He and Guo Chen

Forthcoming:

44. **Gender, Development and Transnational Feminism**
    Edited by Ann M. Oberhauser and Ibipo Johnston-Anumonwo

45. **Fieldwork in the Global South**
    Ethical challenges and dilemmas
    Edited by Jenny Lunn

Cities and Low Carbon Transitions

Edited by
Harriet Bulkeley,
Vanesa Castán Broto, Mike Hodson
and Simon Marvin

LONDON AND NEW YORK

Revised paperback edition published 2013
First published 2011
by Routledge
2 Park Square, Milton Park, Abingdon, Oxon OX14 4RN

Simultaneously published in the USA and Canada
by Routledge
711 Third Avenue, New York, NY 10017

Routledge is an imprint of the Taylor & Francis Group, an informa business

British Library Cataloguing in Publication Data
A catalogue record for this book is available from the British Library

Library of Congress Cataloging-in-Publication Data
Cities and low carbon transitions / Edited by Harriet Bulkeley, Vanesa Castán Broto, Mike Hodson and Simon Marvin. — 1st Edition.
pages cm
Includes bibliographical references and index.
1. City planning—Environmental aspects. 2. Urban ecology (Biology)
3. Climatic changes. 4. Carbon dioxide—Environmental aspects.
I. Bulkeley, Harriet, 1972–
HT166.C466 2013
307.1′216—dc23 2012032000

ISBN: 978–0–415–58697–9 (hbk)
ISBN: 978–0–415–81475–1 (pbk)
ISBN: 978–0–203–83924–9 (ebk)

Typeset in Times New Roman
by Keystroke, Station Road, Codsall, Wolverhampton

Printed and bound in the United States of America by Publishers Graphics, LLC on sustainably sourced paper.

Contents

Illustrations

Figures

Tables

Boxes

Notes on contributors

Alex Aylett is a PhD candidate in the Department of Geography at the University of British Columbia and a Senior Research Associate with the Vancouver-based Sustainable Cities PLUSnetwork. His research deals with the socio-political and economic dynamics of integrated urban climate change planning. He has a particular interest in green urban governance, organisational learning, deep public participation and economic development planning. He has researched urban sustainability in South Africa, Senegal, the United States and Canada. He is a recipient of both SSHRC and Trudeau Foundation research scholarships and has worked as a sustainability consultant for municipalities in Canada and South Africa. He is also a regular contributor to WorldChanging.com.

Harriet Bulkeley is a Professor at the Department of Geography, and Deputy Director of Durham Energy Institute, Durham University. Her research interests are in the nature and politics of environmental governance, with a specific focus on issues of climate change and of urban sustainability. She is co-author with Michele Betsill of *Cities and Climate Change* (Routledge, 2003) and with Peter Newell of *Governing Climate Change* (Routledge, 2010). She is an editor of *Environment and Planning C* and editor of *Policy and Governance* for WIREs Climate Change. She currently holds a Climate Change Leadership Fellowship, 'Urban Transitions: climate change, global cities and the transformation of socio-technical systems', from the United Kingdom's Economic and Social Research Council and co-ordinates the Leverhulme International Network Transnational Climate Change Governance. In 2007, Harriet was awarded a Philip Leverhulme Prize in recognition of her research in this field, under which she is engaged in a project examining the governing of climate change beyond the state in the United Kingdom.

Vanesa Castán Broto is a Lecturer at the Faculty of the Built Environment, University College London. Prior to her appointment she was a Research Associate at the Department of Geography, Durham University, working with Harriet Bulkeley on an ESRC Climate Change Leadership Fellowship, 'Urban Transitions: climate change, global cities and the transformation of socio-technical systems' (2008–2012). Her research focuses on how technology and environmental knowledge mediate the relationship between society and the environment and how this, in turn, influences decision making in urban and regional planning. Prior to working in Durham, she completed an engineering doctorate from the University of Surrey about the deployment of environmental knowledge in the context of conflicts about coal ash disposal and land regeneration in peri-urban areas in Bosnia and Herzegovina. Vanesa has a degree in engineering from the Universidad Politécnica de Madrid (Spain) and an MSc in environmental sociology from Wageningen University. She has published articles and reports for both academic and policy audiences.

Olivier Coutard is a socio-economist. He trained as a civil engineer and holds a PhD in economics and social sciences (1994). He has held a full-time research position with the French National Centre for Scientific Research (CNRS) since 1996. He researches the governance of urban infrastructure services (especially energy and water supply services) and the social and spatial implications of reforms in those sectors. He has been Director of LATTS (http://latts.cnrs.fr) since 2008 and was Director of the French national interdisciplinary research programme on Cities and the Environment (www.pirve.fr), jointly sponsored by CNRS and the French Ministry for Sustainable Development from 2006 to 2009. He is co-editor of *Flux*, an international quarterly on networks and territories (www.cairn.info/revue-flux.htm). He edited *The Governance of Large Technical Systems* (Routledge, 1999) and co-edited, with Richard Hanley and Rae Zimmermann, *Sustaining Urban Networks* (Routledge, 2005).

Shobhakar Dhakal is one of the two Executive Directors of the Global Carbon Project, an international scientific programme based at the National Institute for Environmental Studies, Japan. He largely works on urban carbon and energy modelling for policy applications, low carbon urban scenarios, urban carbon footprints and city climate policies. He was a guest editor for the journal *Energy Policy*'s special issue on Cities and CO_2 emissions, and has published widely on the topic. He contributed to the Global Energy Assessment Consensus Panel on Low Carbon Cities of the Academy of Sciences of South Africa, the city energy modelling expert group of the International Energy Agency, the Task Force on Urban Development and Energy Efficiency of China Council and others. Dr Dhakal was a Senior Policy Researcher and Project Manager of the Urban Environmental Management Project of the Institute for Global Environmental Strategies Japan in the past and holds a PhD from the University of Tokyo and a Master's degree from the Asian Institute of Technology, Pathumthani, Thailand.

James Evans lectures in environmental governance in the School of Environment and Development at the University of Manchester. His research applies theories of knowledge production from science studies and political ecology to the field of urban regeneration. He has over thirty publications, including sixteen research papers in top journals since 2004 and a best-selling book in the field. Currently he is working on the emergence of experimental approaches to urban sustainability, focusing on urban laboratories as drivers of change.

Frank Geels is Professor at SPRU (Science Policy Research Unit) at the University of Sussex, Brighton. He is one of the internationally leading scholars on socio-technical transitions, has co-developed the multilevel perspective and strategic niche management, performed a dozen historical case studies, and applied lessons from these studies to explorations of future transitions using a newly developed socio-technical scenario methodology. His work is interdisciplinary and mobilises insights from science and technology studies, evolutionary economics, history of technology, (neo-)institutional theory and sociology. In 2001, he won the Forbes Prize from the Foundation for the History of Technology for the best junior scholar's publication in the history and sociology of technology. In 2008, he received the Research Publication Award from IAMOT (the International Association for the Management of Technology) for his publications between 2002 and 2007. He recently won a prestigious four-year grant from the European Research Council.

Mike Hodson is Associate Director and Senior Research Fellow at the SURF Centre, University of Salford. His research interests focus on urban and regional transitions to low carbon economies, the ways in which such transitions may or may not happen and understandings of the lessons to be learned from such processes. He has developed projects funded by the European Commission, UK research councils and sub-national government, and through private consultancy. These have principally addressed the relationships between sub-national territories and the reconfiguration of their key socio-technical infrastructures in a period of globalisation and neo-liberalisation, and in a context of the challenges posed by climate change and resource constraint. Mike has published and presented on various aspects of this emerging agenda.

Andrew Karvonen is the Walsingham Capper Research Associate in the Manchester Architecture Research Centre at the University of Manchester. He received a PhD in community and regional planning from the University of Texas at Austin and has over a decade of experience as an urban practitioner in the United States. Through his academic research and teaching activities, he combines theories and approaches from the design disciplines, urban studies, and science and technology studies to explore socio-technical aspects of cities. He is particularly interested in the political implications of sustainable urban development and the reconfiguration of the relationships between technology, nature and humans.

Anne Maassen is a PhD researcher at the Department of Geography, Durham University. Her PhD thesis, titled 'Solar cities in Europe', is a comparative study of solar photovoltaic technology in European cities, with a particular focus on Barcelona, London and Paris. In her research, Anne develops a framework for understanding the place-specific emergence of urban photovoltaics and the impacts that such a technology has on the places of which it becomes a part. Prior to beginning her PhD research in 2007, Anne completed an MSc in Environmental Monitoring, Management and Modelling at King's College

London and a BA in Economics, Politics and International Studies at the University of Warwick. Anne was a fellow at the Institute for Advanced Study on Science, Technology and Society (IAS-STS) in Graz, Austria, for the period October 2009 to March 2010.

Simon Marvin is Carillion Chair of Low Carbon Cities, Professor of Geography and Deputy Director of Durham Energy Institute, Durham University. He is an expert on the changing relations between neighbourhoods, cities, regions and infrastructure networks in a period of resource constraint, institutional restructuring and climate change. Simon's research has been funded by the ESRC, EPSRC, international research foundations, the European Commission, commercial funders and many public agencies. His recent research has addressed comparative urban responses to economic and ecological pressures, in particular in London, New York and San Francisco, and he has also undertaken a placement with Arup on the issues involved in retrofitting cities' infrastructures in response to climate change and resource constraint. Simon has co-authored and co-edited six books on the relationship between cities and infrastructure.

Jenny Pickerill is a Senior Lecturer in human geography at Leicester University. She has published three books and more than twenty academic articles. Her work focuses upon the interrelationships between environment, society and technology. It explores the rhetoric, aims, practices and outcomes of the quest for environmental protection and social justice. In particular, she has researched and written about a variety of solutions to environmental problems – instigated through creative activism, environmental justice projects, diverse approaches to transition and green building. Her most recent project explores the possibilities and implications of affordable eco-housing and is chronicled on her Green Building Blog (http://naturalbuild.wordpress.com).

Harald Rohracher was the co-founder (1988) and Director (1999–2008) of the Inter-University Research Centre for Technology, Work and Culture (IFZ) in Graz, Austria, and is Associate Professor at the Department for Science and Technology Studies, University of Klagenfurt. He has a background in physics (Graz University of Technology), sociology (the University of Graz) and science and technology policy (SPRU, University of Sussex, Brighton). In 2009–2010 he has been Joseph A. Schumpeter Fellow at the Weatherhead Center for International Affairs, Harvard University. In his research, Harald Rohracher is interested in a better understanding of the co-evolution of technology and society and the governance of socio-technical change towards greater sustainability, particularly in the field of energy and the built environment.

Jonathan Rutherford is a Researcher at LATTS (Laboratoire Techniques, Territoires et Sociétés), Université de Paris-Est, where he has worked since 2004. Prior to this, he was a Research Associate at CURDS (the Centre for Urban and Regional Development Studies), Newcastle University, having received his PhD in planning from Newcastle University in 2002. His research explores the conflicting roles of urban infrastructure in the making and remaking of cities and territories in Europe. He has worked on numerous research projects funded by various national and European agencies, and has published widely on the relations between cities and infrastructure networks. He is the author of *A Tale of Two Global Cities: Comparing the Territorialities of Telecommunications Developments in Paris and London* (Ashgate, 2004).

Amanda Smith completed her PhD thesis ('Constructions of "sustainability" and coalfield regeneration policies') in 2005 at the University of Durham in the United Kingdom. The thesis explored the social constructions of sustainability using the coalfield regeneration policies, practices and performances in east Durham as a focus. She worked as a researcher on the ESRC project Regionalisation and the New Politics of Waste at Leeds Metropolitan University before taking her current post at Nottingham Trent University, where she undertakes research and teaching in the areas of urban resilience, sustainable communities and social adaptations to climate change.

Philipp Späth studied political science and geography at the universities of Freiburg and Berlin and obtained his MA from Albert-Ludwigs-Universität Freiburg, Germany. He has worked in the area of renewable energies since 1997, first serving the Environmental Department of the City of Freiburg and then spending four years as the head of a regional association for renewable energy. From 2003 to 2009, Dr Späth was a research fellow at the Inter-University Research Centre for Technology, Work and Culture (IFZ) in Graz, Austria. He received his PhD in science and technology studies with an emphasis on the governance of social-technical change. In December 2009, Dr Späth joined the Institute of Forest and Environmental Policy at the Albert-Ludwigs-Universität Freiburg as an Assistant Professor. His research interests include environmental governance, particularly at regional and local levels; the governance of socio-technical change; the transition of energy systems; and regional organisation of climate change adaptation.

Aidan While is a Senior Lecturer in the Department of Town and Regional Planning at the University of Sheffield. He has long-standing research interests in the field of urban and regional development, with a particular focus on economy–environment relations.

Acknowledgements

We would like to thank the publishing team at Routledge, and in particular Andrew Mould and Faye Leerink, for their support and help with the commissioning and production of this book.

Many of the book's chapters were first presented at the 'Urban Transitions/ Technological Transitions: Cities and Low Carbon Transitions' workshop hosted by SURF in May 2009. This workshop provided a stimulating, interdisciplinary and comparative context for starting to think critically about transitions in an urban context. We were extremely pleased and grateful for the collegiate yet critical style of both research and policy-maker participation in the workshop. Unfortunately, we were not able to publish in this collection all the excellent papers presented at the workshop, but we would like to thank all the participants for their papers and contributions to the debate. We would also like to acknowledge the support for the workshop from the University of Salford and from Harriet Bulkeley's ESRC Climate Change Leadership Fellowship 'Urban Transitions: climate change, global cities and the transformation of socio-technical networks' (Award no. RES-066-27-0002). We would also like to thank Victoria Simpson and Tori Milledge for their assistance in organising the workshop upon which this collection is based, and Tori Milledge for her work in managing the submission of draft chapters.

At SURF, Mike and Simon would also like to thank Tim May and Beth Perry for their continued support in the development of our intellectual interests and political priorities.

Finally, the production of this book has been the product of the editors coming together around a shared set of concerns about both the potential – and the limits – of low carbon transitions in an urban context. It has been a fun and enjoyable process!

Harriet Bulkeley
Vanesa Castán Broto
Mike Hodson
Simon Marvin

1 Introduction

Harriet Bulkeley, Vanesa Castán Broto, Mike Hodson and Simon Marvin

> The transition to a low-carbon economy will be one of the defining issues of the 21st century.
>
> (Secretary of State for Energy and Climate Change, Foreword, *The UK Low Carbon Transition Plan*, 2009)

> We truly don't know if this will work. Transition is a social experiment on a massive scale.
>
> (Transition Network, April 2010)

> We, the mayors and governors of the world's leading cities . . . ask you to recognize that the future of our globe will be won or lost in the cities of the world.
>
> (Copenhagen Climate Change Communiqué, December 2009)

Climate change is a 'wicked problem' (Rittel and Webber 1973). The nature of the climate change problem, in scientific, policy and political terms, is uncertain and there are multiple contested and conflicting discourses concerning how the issue should be addressed. In this context, it has become clear that conventional approaches to environmental management which address single parts of the climate change puzzle one at a time are inadequate. Instead, as the quotations above illustrate, proponents have begun to call for a wholesale *transition* to a low carbon future. In 2009, the UK government published *The UK Low Carbon Transition Plan*, which set out a 'route-map' for 'becoming a low carbon country: cutting emissions, maintaining secure energy supplies, maximising economic opportunities, and protecting the most vulnerable' (DECC 2009: 5). Using similar language but portraying significantly different ideas, the Transition Towns movement, founded in the United Kingdom in 2007 and by 2010 involving more than 250 communities globally, also argues that transition is needed in order to

> look Peak Oil and Climate Change squarely in the eye and address this BIG question: 'for all those aspects of life that this community needs in order to sustain itself and thrive, how do we significantly increase resilience . . . and drastically reduce carbon emissions?'
>
> (Transition Towns 2010)

Low carbon transitions are therefore gaining political salience and public currency. This book seeks to examine the emergence of low carbon transitions and explore their politics and possibilities in the urban arena. The role of cities in addressing climate change has increasingly been recognized over the past two decades (Betsill and Bulkeley 2007; Bulkeley 2010). From being acknowledged by a handful of pioneering municipal authorities in the early 1990s, climate change has risen on the agenda of urban governments and attracted the interest of private and third-sector organizations. Citing the growing proportion of the world's population living in cities in the twenty-first century and their contribution of over 70 per cent of energy-related carbon dioxide emissions (IEA 2008), the mayors of eighty world cities gathered in Copenhagen prior to the 2009 international climate negotiations and called for recognition of the critical role that cities will play in responding to climate change. During the past decade, membership of transnational networks such as the ICLEI Cities for Climate Protection and Climate Alliance has grown, and new networks, including the C40 Cities Climate Leadership Group and the Rockefeller Asian Cities Resilience Network, have been formed. In the United States, in 2005 the mayor of Seattle, Greg Nickels, challenged mayors across the nation to take action on the issue, and by 2009 over 900 mayors had signed up to the US Mayors' Climate Protection Agreement (Gore and Robinson 2009: 143). This approach has been replicated globally, most recently with the launch in 2009 of the European Covenant of Mayors, which now has more than a thousand members. At the same time, private actors, including financial institutions, property development companies, utilities, foundations and non-governmental organizations, are increasingly involved in initiatives to address climate change. Further, new grassroots networks, including 'transition towns', are emerging which take the urban as an explicit arena within which to address climate change.

The conjunction of the growing prominence and plurality of urban climate change responses and the emergence of calls for a low carbon transition raises important questions about the future of cities, and in particular the infrastructures that sustain urban life. The provision and organization of urban infrastructures – including energy, water, waste, shelter and mobility – have largely been perceived as unproblematic and taken for granted as primarily engineering challenges and administrative issues (Graham and Marvin 2001). Recent analyses have called these assumptions into question, demonstrating the critical role played by infrastructures in urbanization and vice versa, and highlighting structural shifts taking place in the provision and use of urban networks. Recognizing the intimate connection between, on the one hand, cities and climate change and, on the other hand, urbanization and infrastructure systems, it is increasingly clear that addressing climate change will require fundamental transformations in the urban infrastructure networks that sustain daily life. In short, it is clear that urban infrastructure networks will be central to any effort to achieve a low carbon transition.

There is therefore a pressing need to consider how system transitions take place within the city. Such systems, we suggest, should be conceived as *socio-technical* – that is, they comprise, and are co-produced by, social and technical elements. For example, a photovoltaic energy system comprises a form of energy conversion technology (photovoltaic cells), made from materials (e.g. silicon), installed through a particular configuration of technical artefacts (e.g. a building integrated system), in the context of political and legal institutions (e.g. planning requirements), processes of design (e.g. house building) and social practices (e.g. domestic use of electricity). The resulting system is considered socio-technical because it emerges through the conjunction and co-evolution of these 'technical' and 'social' entities and processes. While there are a variety of approaches to understanding socio-technical systems and their dynamics, the one which has to date engaged most explicitly with the challenges of sustainability and in particular energy systems is that based on the 'multilevel' analysis of systems in transition (Geels 2004; Elzen *et al.* 2004). However, while there is a growing scholarship on transitions in socio-technical systems, this work has to date paid little attention to the urban scale. In this book, we critically examine this underexplored relationship between the multilevel perspective (MLP) on socio-technical transitions and cities, where pressures to undertake low carbon transitions in infrastructure systems are, as we have outlined above, particularly pronounced. In so doing, we seek to answer the key question: how, why and with what implications are cities effecting low carbon transitions?

In order to address this question, the book is organized in two parts. Part I examines the key interconnections between cities and transitions by looking critically at the historical, theoretical, conceptual and methodological relations between the urban, infrastructural systems and socio-technical change. It outlines the key concepts underpinning theories of socio-technical transition, the emergence of the low carbon imperative within cities, and the extent to which such a phenomenon might be understood in terms of systemic change. Part II is more empirical in orientation. Through a series of case studies, contributors address how, and with what consequences, we understand the role of 'the city' in undertaking urban transitions in practice. In the remainder of this Introduction, we consider in turn these two related aspects of the book in more detail before introducing the contributions made by each of the following chapters.

Conceptualizing low carbon transitions and the urban

As the quotations at the beginning of this Introduction make clear, there are many different ways in which the concept of a 'transition' might be understood. In seeking to understand the nature and potential of urban low carbon transitions, Part I of this book draws on three sets of debates to outline some of the key issues concerning how we might theorize such processes. First, as we set out above, a primary concern of this volume is to consider the applicability to the urban arena of insights from the debates concerning socio-technical regimes and their transition (Geels 2002; Shove and Walker 2007; Smith *et al.* 2010; Hodson and Marvin 2009). As a field of study that has explicitly engaged with the ways in which Large Technical Systems, such as energy systems, remain stable and undergo change, this analytical framework holds significant promise for understanding the ways in which urban infrastructure networks may be transformed in response to climate change. Central to current analyses of the historical and future transformation of socio-technical systems is the 'multilevel perspective', which suggests that such systems can be analysed in terms of a broad landscape of institutions and norms, distinct socio-technical regimes that structure the ways in which particular systems operate and their dynamics, and niches where innovation and experimentation take place (Geels and Kemp 2007; Geels and Schot 2007; Smith *et al.* 2010). Transformation in such systems can be both incremental and radical, and is dependent on the alignment of innovations at the niche level, with windows of opportunity created within incumbent regimes and through the dynamics of landscape change. The appeal of such an analytical framework lies both in its comprehensive nature and in its ability to explain long-term and far-reaching shifts in socio-technical systems. However, critics have cautioned that the politics of transitions receive limited attention in a framework more concerned with innovation and social learning (Shove and Walker 2007), and that the places within and through which transitions occur have largely been absent from analysis (Hodson and Marvin 2010). In Part I of this book, the authors consider the merits and limitations of the multilevel perspective and in particular seek to bring concepts of place and the urban into the analytical frame, while in Part II the contributors consider the analytical purchase of the multilevel perspective for explaining urban responses to the low carbon imperative.

A second set of debates with which this book directly engages considers urban low carbon futures in terms of policy, politics and governance. Within the growing literature on urban responses to climate change, analysis has predominantly focused on the processes of policy change which have or have not sustained shifts towards low carbon strategies and measures in the urban arena. The extent of policy transformation is explained in terms of issues of institutional capacity and urban politics, and conceived as structured within a 'multilevel governance' framework where decisions and actions at international, national and regional levels and emerging through governance networks serve to create the conditions of possibility within individual cities. While there are similarities here with the MLP on socio-technical systems transition, in terms of recognizing the different arenas through which transitions are simultaneously produced, in the multilevel governance approach these arenas are regarded not as separate entities but as continually structuring and reproducing one another. However, analyses of policy change have rarely taken into account the socio-technical, or explicitly addressed questions of transition. Chapters in this book take up these challenges, examining the growing imperative of 'carbon control' in urban governance and the emerging political economies of low carbon transitions (Bulkeley *et al.*, Chapter 3; Hodson and Marvin, Chapter 5; While, Chapter 4). In these accounts, cities are reproduced through, and at the same time actively reconfigure, structural social, economic and political processes. Cities become critical to 'state strategies' of carbon control, while urban actors respond to new low carbon imperatives in a variety of ways, from radical plans for localization to efforts to embed urban economies in low carbon technological futures, in turn reshaping what effective and legitimate state strategy might entail. In short, such analyses suggest that the urban is a fundamental part of low carbon transitions, whatever their spatial scale.

The third set of debates with which this collection engages problematizes the relation between the urban and infrastructure systems further through focusing on processes of 'urban metabolism', in which the city is described as a process of socio-ecological change (Swyngedouw and Heynen 2003). Rather than attending to the political economy of socio-technical systems, these debates are concerned with the ways in which infrastructure (and other) networks mediate socio-ecological flows in the city. The resulting political ecologies consider how uneven power relations orchestrate and are reproduced by the city, examining human and non-human agency in these processes. The flux and flow of urban metabolism suggests that transformation is a continual process, and, rather than being taken for granted, the fixity of urban socio-technical systems is a complex accomplishment. In this reading, transitions in urban systems are as likely to emerge from the co-incident actions of multiple agents and everyday actions as from purposive attempts to transform the city. Indeed, the success, or otherwise, of transitions may be located not in broader political and economic processes, or in terms of the processes of innovation, but through the ways in which they are mediated by everyday life and the myriad power relations that sustain and constrain such actions. Contributors in Part I of this book consider the implications of this alternative understanding of the urban, and of its transformation, suggesting that it may open up the analytical space for considering transitions in more diverse and plural forms while placing issues of social and environmental justice at the centre of the research agenda (Bulkeley *et al.*, Chapter 3).

Cities in transition?

If Part I is more conceptual in orientation, Part II of this book starts the significant and pressing task of understanding the processes of low carbon transition that are taking place in cities. Drawing from a range of urban places – from global cities to local communities, in the North and global South – the contributors focus on three sets of related issues. The first, and in some ways most fundamental, concerns whether we can identify processes of 'low carbon transition' taking place within urban arenas and, if so, what they entail. Contributors take diverse methodological approaches to this question, for example, considering the changing nature of energy use and demand (Dhakal, Chapter 6), providing an historical analysis of the emergence and implementation of low carbon strategies and measures (Späth and Rohracher, Chapter 7), or demonstrating the 'performance' of low carbon transitions in the Transition Town movement (Smith, Chapter 11). Nonetheless, across a range of cases the chapters in Part II suggest that the imperative of responding to climate change is firmly on the urban agenda. Evidence is found for the emergence of transitions – in energy use (Dhakal, Chapter 6), in low carbon urban planning and development (Evans and Karvonen, Chapter 9; Pickerill, Chapter 12; Späth and Rohracher, Chapter 7) and urban infrastructure systems (Aylett, Chapter 10; Coutard and Rutherford, Chapter 8) – but the picture is fragmented. While, following the multilevel perspective on transitions, this may be interpreted in terms of the emergence of low carbon urban niches, the contributors also find theoretical insights from the literatures on urban infrastructures, political ecologies and policy useful in explaining their findings.

A second set of issues that Part II addresses concerns how transitions are taking place and who is involved in these processes. For many of the contributors, in contrast perhaps to the relatively comprehensive accounts of historical transitions presented in the literature on Large Technical Systems and socio-technical regimes, urban low carbon transitions appear to be fluid, contested and partial processes. The one exception may be the case of urban energy use in China, where the national government's policies for development and industrialization have steered urban energy trajectories towards greater energy consumption. In the other contributions, low carbon transitions are regarded as driven either through processes of urban governance, broadly defined, or through some form of social movement. First, contributors show how the fragmentation of urban infrastructure systems, the changing nature of urban governance, the creation of visions or shared discourses around urban futures, and the formation of coalitions across public/private boundaries and institutional structures have shaped the opportunities for reframing urban development processes in low carbon terms. Second, the chapters demonstrate that low carbon transitions are not necessarily the preserve of an urban elite; they are also being driven by social movements and individual actors. Here too, shared visions, creating coalitions and networks, and operating across the public/private divide are also seen as important. Cities are therefore involved in shaping and directing transitions, but the capacity to do so, the actors involved and the politics of these processes vary from place to place.

The third set of issues that the chapters in Part II raise is the extent to which urban places are shaping the conditions of possibility for low carbon transitions – or, in other words, what difference does the urban make? The chapters show how cities have provided both political opportunities – for example, coalition building – and openings for particular kinds of socio-technical intervention, such as combined heat and power plants and solar photovoltaic cells (Evans and Karvonen, Chapter 9; Coutard and Rutherford, Chapter 8; Späth and Rohracher, Chapter 7). At the same time, some contributors point to the ways in which making transitions in the urban arena is more challenging than in smaller communities or in rural settings, because of the challenges of developing in innovative ways within the existing urban fabric (Pickerill, Chapter 12; Smith, Chapter 11), and the limiting effects of existing institutional and political structures (Aylett, Chapter 10). Collectively, they suggest that the urban is critical in both enabling and constraining potential low carbon transitions, but at the same time point to the importance of regional and national government, transnational networks, private-sector actors operating locally and/or globally, and external events in shaping their trajectories.

Outline

Having outlined the main debates considered in this book, we turn here to the contributions of each chapter in turn. Following this Introduction, in Chapter 2 Geels provides an overview of the 'multilevel perspective' on socio-technical transitions and discusses the relevance of different scales of social and political organization within this framework. While analyses using the multilevel perspective have focused on examples of transitions occurring at the national level, Geels suggests that transitions can also be studied at other levels (e.g. international, local/city), depending on the issues and audience in question. Focusing on the historical analyses of national transitions, Geels argues that a city can play three roles: as a primary actor leading the transition; as a seedbed of innovations which may gather pace through the creation of national infrastructure; or as having a limited role. The chapter provides a coherent analysis of the potential for the multilevel perspective in the urban context.

In Chapter 3, Bulkeley *et al.* outline how the city has come to prominence in the climate change debate. They argue that rather than regarding the urban climate change agenda in policy terms, there is a need to look to its socio-technical nature. Rather than regarding the city as either an actor or an arena within which transitions take root, they suggest that an engagement with the social studies of urban infrastructure systems shows the city as a complex web of social and ecological processes through which networks are sustained and contested. These perspectives, they argue, are particularly useful in revealing the ways in which obduracy and flux in urban systems are created, maintained and contested, and in enabling an analysis of the political implications of urban transitions. In Chapter 4, While also argues for the importance of theorizing the urban politics of transitions, making the case for an emerging 'carbon calculus' as critical to contemporary urban governance. In While's analysis, urban transitions take place at the intersections between the reconfiguration of state strategy around the political economy of carbon and the restructuring of urban infrastructure systems. By recasting the city as a space of carbon flows, new forms of intervention and practice become desirable, legitimate and even necessary. In this manner, the city is not separate from transitions, but fundamental to their emergence and development. In Chapter 5, Hodson and Marvin address two questions: Can cities shape socio-technical transitions? And how would we know if they were doing so? In addressing these questions, they set out emerging evidence that purposive and managed change in the socio-technical organization of infrastructure networks – characterized as 'systemic' transition – is unfolding in the context of world cities. Hodson and Marvin use these developments as an entry point to develop a conceptualization of the role of 'the city' in undertaking transitions and a review of the strengths and shortcomings of the MLP in addressing this. The chapter uses this engagement to identify what an urban transition would look like, and constructs a new framework to conceptualize and research urban transitions.

In Part II of the book, Dhakal in Chapter 6 provides an historical account of the emerging 'high carbon' transition taking place among Chinese cities in response to the drive for development and industrialization. Unpacking this phenomenon, the chapter demonstrates the place-specific nature of this transition, and the ways in which counter-currents towards low-carbon urban development are emerging in some of the largest Chinese cities, focusing on Shanghai as a case study. In Chapter 7, Späth and Rohracher also provide an historical analysis of energy transitions, although through a governance lens. They analyse the transition to sustainability in Freiburg, Germany, and Graz, Austria, and show how new forms of discourse and governance are critical in shaping low carbon transitions. Their analysis demonstrates that different actors drive the transitions, whether led by politicians and administrators, as in Freiburg, or by entrepreneurs and knowledge makers, as in Graz, and that the potential for urban innovations to have a wider impact is conditioned by the 'multilevel' governance context within which they are situated.

In Chapter 8, Coutard and Rutherford examine the contemporary urban scene, arguing that the rise of new forms of experimentation with decentralized and localized urban infrastructure networks can be interpreted as part of the development of a 'post-networked city'. Drawing on the 'splintering urbanism' thesis and comparing its potential with theories of socio-technical transition, they argue for an approach that finds a middle ground between these alternatives. Identifying four different pathways through which the transition to a post-networked urban condition is emerging, and illustrating these with examples from across Europe, they use this framework to demonstrate the potential transformative power of urban experimentation but also the political ambivalence of the post-networked city. In Chapter 9, Evans and Karvonen point to the ways in which 'living laboratories' are becoming a means through which the development of, testing of and explicit self-reflection on 'niche' urban infrastructures are being undertaken. Often orchestrated through universities and with the explicit target of creating new forms of knowledge about what in Coutard and Rutherford's terms we might call the 'post-networked' city, they argue that living laboratories are an increasingly important vehicle through which knowledge of the city and plans for its future are being created. Drawing on examples from the Middle East, North America and the United Kingdom, they show how the living laboratory idea is being deployed and consider the potential implications for future urban development. In Chapter 10, Aylett focuses on the case of Durban, South Africa, a city that has experienced significant energy network shocks and lacks many of the advantages of the 'networked' city of the North. In his chapter, Aylett argues that the 'trained incapacity' created through organizational cultures and professional training can prevent experimentation even where opportunities may arise. In so doing, Aylett demonstrates the multiplicity of transitions taking place within one city, the ways in which agency is enabled and constrained by institutional structures, and the implications for how we consider the process and outcomes of transitions.

Chapter 11, by Smith, documents the emergence of the Transition Town movement, which emerged in the United Kingdom but has since spread to many countries of the OECD. Analysing the 'performative discourses' involved in Transition Towns, using Nottingham in the United Kingdom as an example, Smith examines the process of transition as a social one where different actors struggle to put forward their visions of what a transition entails and how it should be achieved. Although the movement has so far been successful in terms of developing the case for a transition, as this comes to be undertaken in practice, Smith questions the capacity of the movement to create momentum and catalyse a transition to a low carbon system. In Chapter 12, Pickerill also conceives of the process of transition through the lens of social movements. She explores the opportunities for Low Impact Developments to act as innovations and catalyse change in the urban setting, drawing on examples from the United Kingdom and Germany, and documents cases of success and failure. Through this analysis, she argues that Low Impact Developments are innovations that seek to accomplish change from the 'margins'. This perspective opens up a critique of transitions as a 'blueprint', pointing instead at transitions occurring on a multitude of levels, and the complexity of 'engaging in radical social change'.

Individually and collectively, these chapters make a considerable contribution to how we can understand the emerging phenomenon of urban low carbon transitions. In Chapter 13, we revisit the key themes raised in this Introduction and throughout the contributions to the book to offer some conclusions about the state of the field to date and the directions for future research.

References

Betsill, M. and Bulkeley, H. (2007) 'Looking back and thinking ahead: A decade of cities and climate change research', *Local Environment: The International Journal of Justice and Sustainability* 12 (5): 447–456.

Bulkeley, H. (2010) 'Cities and the governing of climate change', *Annual Review of Environment and Resources* 35 (forthcoming).

Bulkeley, H. and Betsill, M. M. (2003) *Cities and Climate Change: Urban Sustainability and Global Environmental Governance*, London: Routledge.

Copenhagen Climate Communiqué (2009) Copenhagen Climate Communiqué, Copenhagen 2009. Available at: www.kk.dk/Nyheder/2009/December/~/media/B5A397DC695C409983462723E31C995E.ashx (accessed May 2010).

DECC (2009) *The UK Low Carbon Transition Plan: National Strategy for Climate and Energy*, London: Department for Energy and Climate Change.

Elzen, B., Geels, F. and Green, K. (eds) (2004) *System Innovation and the Transition to Sustainability: Theory, Evidence and Policy*, Cheltenham, UK: Edward Elgar.

Geels, F. W. (2002) 'Technological transitions as evolutionary reconfiguration processes: A multi-level perspective and a case-study', *Research Policy* 31 (8–9): 1257–1274.

Geels, F. W. (2004) 'From sectoral systems of innovation to socio-technical systems: Insights about dynamics and change from sociology and institutional theory', *Research Policy* 33 (6–7): 897–920.

Geels, F. W. and Kemp, R. (2007) 'Dynamics in socio-technical systems: Typology of change processes and contrasting case studies', *Technology in Society* 29 (4): 441–455.

Geels, F. W. and Schot, J. (2007) 'Typology of sociotechnical transition pathways', *Research Policy* 36 (3): 399–417.

Gore, C. and Robinson, P. (2009) 'Local government response to climate change: Our last, best hope?' In H. Selin and S. D. VanDeveer (eds) *Changing Climates in North American Politics: Institutions, Policymaking and Multilevel Governance*, Cambridge, MA: MIT Press.

Graham, S. and Marvin, S. (2001) *Splintering Urbanism: Networked Infrastructures, Technological Mobilities and the Urban Condition*, London: Routledge.

Hodson, M. and Marvin, S. (2009) 'Cities mediating technological transitions: Understanding visions, intermediation and consequences', *Technology Analysis and Strategic Management* 21 (4): 515–534.

Hodson, M. and Marvin, S. (2010) 'Can cities shape socio-technical transitions and how would we know if they were?', *Research Policy* 39: 477–485.

IEA (2008) *World Energy Outlook 2008*, Paris: International Energy Agency.

Rittel, H. and Webber, H. (1973) 'Dilemmas in a general theory of planning', *Policy Sciences* 4 (2): 155–169

Shove, E. and Walker, G. (2007) 'CAUTION! Transitions ahead: Politics, practice and sustainable transition management', *Environment and Planning A* 39: 763–770.

Smith, A., Voss, J. and Grin, J. (2010) 'Innovation studies and sustainability transitions: The allure of the multi-level perspective and its challenges', *Research Policy* 39, 435–448.

Swyngedouw, E. and Heynen, N. C. (2003) 'Urban political ecology, justice and the politics of scale', *Antipode* 35 (5): 898–918.

Transition Network (2010) 'Welcome'. Online, available at: www.transitionnetwork.org/ (accessed May 2010).

Transition Towns (2010) Homepage. Online, available at: www.transitiontowns.org (accessed May 2010).

Part I

Conceptual frameworks for understanding urban transitions

2 The role of cities in technological transitions

Analytical clarifications and historical examples

Frank Geels

Introduction

This chapter works from the field of transition studies to draw out lessons with regard to the role of cities in technological transitions at the *national* level. The chapter describes the multilevel perspective (MLP) on transitions, introduces some analytical distinctions with regard to the role of cities, provides historical examples as illustrations of these roles, and briefly discusses the implications for low carbon transitions. But first it introduces the topic of transitions and system innovations, and positions the socio-technical approach in relation to established ways of thinking.

New environmental problems, such as climate change, biodiversity and resource depletion, gained prominence on the political agenda in the 1990s and early 2000s. These pervasive problems differ in scale and complexity from the environmental problems of the 1970s and 1980s, such as water pollution, acid rain, local air pollution and waste problems. While many of these problems could be addressed with end-of-pipe solutions (e.g. catalysts in cars, scrubbers on power stations) or clean technologies, new environmental problems such as climate change are more difficult to address and will require social as well as technical changes. Achieving cuts of 50–80 per cent in CO_2 emissions will require shifts to new kinds of systems in transport, energy and agri-food domains. Such transitions entail not only new technologies but also changes in markets, user practices, infrastructures, cultural discourses, policies and governing institutions. The dynamic interactions and co-evolution between these elements have been investigated in the literature on socio-technical transitions (Kemp 1994; Geels 2002, 2004; Elzen *et al.* 2004; Wiskerke and Van der Ploeg 2004; Smith *et al.* 2005; Shackley and Green 2007; Geels and Schot 2007; Verbong and Geels 2007). The socio-technical approach focuses on *multiple* actors and social groups, not only firms or consumers and markets. The multidimensionality of socio-technical change implies that transitions are complex multi-actor processes that often take several decades to come about.

A widely used example of this socio-technical approach is the multilevel perspective (MLP), which is briefly discussed in the next section. I have tested and refined the MLP with a dozen historical case studies in various domains. These studies, and most other work on socio-technical transitions, have focused on the *national* systems level[1] because cultures, infrastructures, regulations and institutions are mostly (but not always) national phenomena, leading to different national *styles* in transport, food and energy systems. But one can also study transitions at international or urban levels. Because transport and energy systems in different countries share many characteristics (e.g. similar kinds of technologies, behavioural patterns, the same multinational industries operating in different countries), one could argue that *international systems* are the right level of analysis. But one can also focus on differences *within* national systems. The timing, speed and causal mechanisms of transitions in various cities may differ, for example. Some examples of transitions that have been studied at the city level are:

- the transition from coal to gas in Pittsburgh, 1940–1950 (Tarr 1981);
- the transition from manual to mechanized unloading of ships in the port of Rotterdam, 1890–1920 (Van Driel and Schot 2005);
- the transition to piped water systems in Amsterdam, 1850–1920 (Groen 1980).

These studies tend to focus on local specificities (e.g. geographical conditions), visionary individuals (e.g. mayors, reformers), politics and power struggles (e.g. in city councils), and pressures from reform coalitions (e.g. urban interest groups, chambers of commerce).

So, there are reasons why the transitions literature so far has focused on the national level. But there are also arguments for studying transitions at other scale levels (international, local/city). The ‘best’ level of analysis cannot, of course, be specified beforehand, but depends on the debate, research questions and the audience scholars aim to address.

The chapter discusses the roles cities play in *national* transitions. This discussion is organized around an analytical distinction that proposes three kinds of roles:

1. Cities and city governments as primary actors of national transitions. This role is likely when national systems/regimes are collections of local systems.
2. Cities as seedbeds of national transitions. These transitions may start in cities, but subsequently gather pace through the creation of ever-larger infrastructures, resulting in national systems. Cities are the location or ‘space’ where transitions start. But city governments are only one actor, others being consumers, firms, and so on.
3. Limited role of cities. For transitions in nationally operating systems (often related to national consumer markets), the role of cities may be limited.

Later sections will elaborate on this analytical distinction and use historical case studies to provide illustrations and highlight relevant dynamic mechanisms. The chapter’s final section draws conclusions and extends the argument to low carbon transitions. First, however, the multilevel perspective on transitions is briefly introduced in the following section.

Multilevel perspective on transitions

An important embodiment of the socio-technical approach is the multilevel perspective, which can be only briefly described here (for further elaboration, see Rip and Kemp 1998; Geels 2002, 2004, 2005, 2006; Geels and Schot 2007). The MLP distinguishes three analytical levels: niches (the locus for radical innovations); socio-technical regimes, which are locked in and stabilized on several dimensions; and an exogenous socio-technical landscape. These 'levels' refer to heterogeneous configurations of increasing stability. To help us to understand system transitions, the MLP analyses interacting processes within and between these levels (see Figure 2.1). Socio-technical transitions come about through multidimensional alignments of processes within and between these three levels. (1) Niche innovations build up internal momentum through co-construction of heterogeneous elements in a stable configuration, through learning processes and price/performance improvements, and through support from powerful groups. (2) Changes at the landscape level

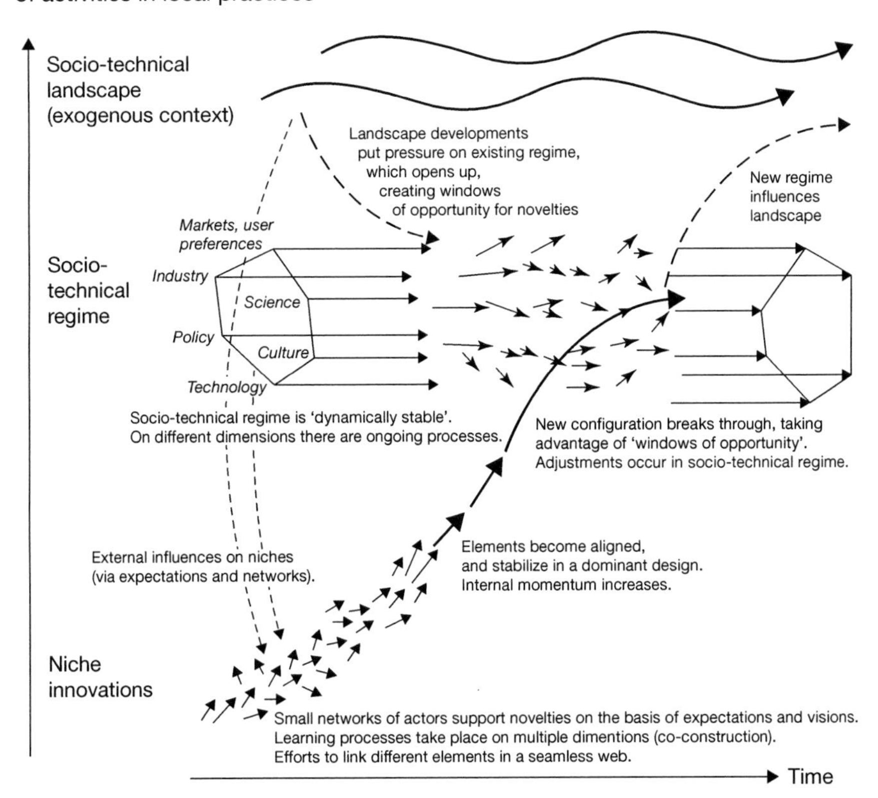

Figure 2.1 Multilevel perspective on transitions.

create pressure on the regime. (3) Destabilization of the regime creates windows of opportunity for niche innovations. The alignment of these processes enables the breakthrough of novelties in mainstream markets where they compete on multiple dimensions with the existing regime.

The MLP has an enactment logic *within* the levels (inspired by social constructivism and structuration theory), with trajectories and lineages resulting from social (inter)actions within semi-coherent rule structures that are recursively reproduced and incrementally adjusted by ongoing actions (Geels 2004). And it has an evolutionary logic *between* the levels, with niche innovations providing (radical) variety that interacts with broader selection environments (at regime and landscape levels). Selection is seen as a multidimensional process that not only involves market processes, but also depends on the fit of niche innovations with regulations, infrastructure and cultural meaning.

The socio-technical approach and multilevel perspective differ from three other views on environmental problems, which respectively tend to focus on markets, technology or behavioural change. I discuss these established views to indicate how the socio-technical approach differs from them (based on Geels *et al.* 2008). Because of space constraints, this discussion is brief and leaves out various nuances in these views.

First, neo-classical economics and neo-liberalism view environmental problems as negative externalities resulting from market failures. The suggested policy strategy is internalization of external costs (e.g. through 'polluter pays' schemes or emissions trading) in order to 'get the prices right'. The 'right' prices then create incentives for producers to develop and adopt cleaner technologies, and for consumers to adopt 'greener' products. While this approach may work under certain conditions (rational agents, full information, perfect markets), it is problematic for radical innovations and transitions that are characterized by uncertainties (about technologies, user preferences, market institutions) which complicate rational calculations. Neo-liberal approaches are also less effective when there are no level playing fields on which old and new technologies can compete or when existing systems are stabilized by a lock-in mechanism.

Second, ecological modernization sees environmental problems as a serious threat to modern societies, and suggests a green reorientation of modernism, based on clean technology, eco-efficiency, dematerialization and the closing of material loops. It maintains a belief in modernist principles such as science, technical progress, control and economic growth. Although ecological modernization introduces a welcome 'supply-side' perspective, critics question its sufficiency for sustainable development and argue for more radical changes (York *et al.* 2003).

Third, 'deep ecology' and eco-centrist approaches see environmental problems as a sign of deep failures of modernist values and societies. These values should therefore be replaced by 'green values' and behavioural change (e.g. localism, self-sufficiency). Less radical versions emphasize community-based initiatives where villages or neighbourhoods collectively adopt, maintain and administer 'green' technologies (e.g. biogas plants, solar panels, wind turbines) or collectively stimulate behaviour change. While this approach usefully highlights *social* innovations, its radical overtones and (sometimes) technophobia may restrict its appeal to the broader public.

Compared to these established views, the socio-technical approach suggests that transitions come about through the co-evolution of social and technical elements, and entail interactions between many social groups. Against this background, I now turn to a discussion of the roles cities play in *national* transitions. Three subsequent sections discuss and present historical examples of three roles: (1) cities and city governments as primary actors in national transitions; (2) cities as seedbeds in early phases of national transitions; and (3) cities playing only a limited role.

Cities as primary actors

Some national systems or regimes consist of a collection of urban or local systems of provision. Waste removal, water supply, and trams and light rail systems, for instance, are often operated at local or regional levels. Aggregate regimes in these cases may have national dimensions in terms of regulations or technical knowledge (e.g. embodied in engineering communities), but implementation and operation tend to occur in cities and localities. City governments are crucial actors in managing and organizing these systems. National transitions in these kinds of systems consist of accumulations of local and urban transitions.

The multilevel perspective incorporates this pattern via the implicit 'level' of local practices for which niches, regimes and landscape provide structuration (see the *y*-axis of Figure 2.1). In particular, niches and regimes are always enacted and reproduced by actors who are situated in local practices. Scholars often operationalize local practices in terms of experimental projects (with new technologies, user preferences, infrastructures, regulations) in different localities. For radical innovations, Geels and Raven (2006) for instance, make a distinction between, on the one hand, local innovation projects and, on the other, the niche level, which consists of an emerging community that shares particular beliefs, goals and knowledge (Figure 2.2).

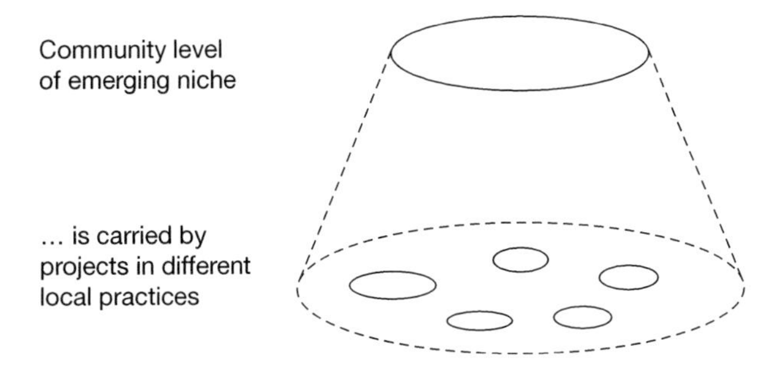

Figure 2.2 Local projects and technological niche.

But for the systems we discussed above, it is also possible to operationalize local practices in terms of different cities and localities. Emerging systems (niche innovations) or existing national systems can then be seen as carried and enacted by different cities. In such a conceptualization, one can further investigate how innovators in different cities organize themselves into broader communities, build networks, exchange knowledge and experiences through workshops and conferences, and articulate best practices. While transitions start haphazardly in a few cities, they may gather momentum when more cities join, when networks become stronger, when experiences stabilize into dominant designs, and when support coalitions are expanded (e.g. through support from national ministries).

An illustration of this role of cities can be found in the historical transition from wells to piped water in the Netherlands between 1870 and 1930 (Geels 2005). Table 2.1 clearly shows that this transition proceeded at different speeds in different cities, with a marked acceleration in the 1880s and 1890s.

The creation of piped water systems generally occurred sooner in larger cities, which had more people with sufficient buying power. Amsterdam (250,000 residents in 1850) was the first city to make the change, in 1853, primarily because water from wells and canals was of low quality and brackish (owing to infiltration of sea water from under the dunes) (Groen 1980). Rotterdam (120,000 residents in 1870), The Hague (100,000 residents in 1870) and Groningen (50,000 residents in 1880) soon followed. An exception to this rule was the small town of Den Helder (15,000 residents in 1850), which introduced piped water systems to supply outgoing merchant ships with clean water.

Table 2.1 Year of implementation of waterworks in Dutch cities

City	Year	City	Year
Amsterdam	1853	Delft, Amsterdam (Vecht water), Niewer Amstel, Kampen, Oud-Bijsterland, Leeuwarden	1888
Den Helder	1856	Venlo, Zutphen	1889
Rotterdam	1874	Tiel, Amersfoort	1890
The Hague	1874	Maassluis	1891
Leiden, Katwijk	1878	Enschede, Almelo, Middelburg	1892
Nijmegen	1879	Zwolle, Almelo, Deventer	1893
Groningen	1881	Breda, Apeldoorn, Meppel, Delden	1894
Dordrecht	1882	Tilburg	1895
Delfshaven, Utrecht, De Bilt	1883	Hellevoetsluis, Harderwijk, Zeist	1896
Vlissingen, Arnhem, Baarn, Soest, Alkmaar, Vlaardingen	1885	Hengelo, Assen	1897
Zaanstreek, Gorinchem,	1886	Nijkerk, Rheden, Zwijndrecht	1898
Schiedam, Hilversum Sliedrecht, 's-Hertogenbosch, Roosendaal, Maastricht	1887	Bergen op Zoom, Helmond, Roermond	1899

Source: Constructed from data in Vogelzang (1956).

The transition towards clean piped water systems is often portrayed as a (rational) response to expanding cities, increasing water pollution (because excrements and urine were released untreated into the environment), advances in medical science (Pasteur's discovery of micro-organisms in the 1860s) and high mortality rates due to infectious water-borne diseases (dysentery, cholera, typhoid). Although hygiene and public health were often discussed (as climate change is nowadays), in-depth studies show that these were not the main motivations for cities, citizens and firms to switch to piped water systems. City governments had locally different reasons to stimulate piped waterworks. In some cities, such as Rotterdam, piped water could piggyback on a canal-flush sewer system which had been created in the early 1870s (Van den Noort 1990). Drinking water was a by-product of the canal-flush sewer system, to help pay for its costs. A second reason to construct piped waterworks could be to raise the city's prestige. The city of Maastricht, for instance, worried about the middle classes moving out, which was a threat to the tax base. To improve the city's prestige, Maastricht implemented waterworks which not only provided clean piped water but also powered many decorative fountains (Cillekes *et al.* 1988). Third, for cities that suffered from quantity problems (e.g. Amsterdam), piped water was an additional means of providing drinking water. A fourth reason was to improve hygiene and public health, but this was seldom the deciding factor (Geels 2005). The first users of piped water were not the poor working classes who might have experienced real health benefits from clean drinking water, but rich people, who usually already had other means of getting clean water (e.g. private pumps, rainwater basins, buying of imported water).

Various user groups (rich citizens, industrial firms, municipal agencies) also had different motivations to pay for piped water. An important reason for rich people was increased comfort and convenience. Piped water offered a new functionality in water supply. It was no longer necessary to go outside (to a pump, river or well); one could simply turn the tap on. Second, piped water was a means by which to show status and social distinction. Third, quantity problems could create demand for fresh and clean drinking water. In Amsterdam and The Hague, rich citizens were willing to pay for clean dune water because it was better than water from local sources. A fourth reason was the linkage of piped water with other sanitary techniques (e.g. baths, showers and washing bowls). Piped water was linked with the culture of cleanliness, facilitating more washing and scrubbing (Geels 2005). Piped water was also used for water closets, which experienced rapid growth in the 1890s.

Another early user group was industrial firms with particular needs for clean water (e.g. paper factories, breweries). For instance, in the city of Tilburg, waterworks supplying piped water were primarily created for the textile industry. In its first year of operation, about 91 per cent of Tilburg's piped water was used by industry and 9 per cent by local residents (Van der Heijden 1995). Local residents could thus piggyback on industrial water demand. Another industrial use of water was for the production of steam for steam engines. Firms were willing to pay for clean water, because free but polluted water might lead to deposits on boilers, reducing their efficiency (Cillekes *et al.* 1988). Other early users were municipal agencies such as fire departments, cleaning departments (which wet the streets to diminish dust problems, and used water to flush out public urinals).

In sum, the shift to piped water systems was not primarily driven by hygienic considerations (about disease and clean drinking water). Instead, there were many motivations, which often differed between various cities. Cities also differed in terms of the organizational and operational dimensions of their piped water systems. In terms of organizational form, the systems could be provided privately or publicly. Early water systems in Amsterdam and Den Helder were private and provided by commercial companies. City governments were initially hesitant to get involved. But because private initiatives could be slow and risk averse, city authorities gradually began considering constructing waterworks themselves. The city of Rotterdam had shown that this was possible in 1874. Subsequently, the number of public initiatives increased, although private waterworks also expanded (Figure 2.3). The coexistence of private and public options created ongoing debate and uncertainty about the best organizational form (Geels 2005).

Cities also differed in terms of the source of water supply. Some cities used groundwater from the dunes (Haarlem, Amsterdam, The Hague); others used water from rivers (Rotterdam, Maastricht) or from small lakes (Tilburg). These differences had implications for the length of pipes, the kinds of pumps needed, the kind of water filtration systems, and overall costs. Additionally, city engineers had to address questions such as: How large and reliable were these sources? What would be the effect of drinking water on groundwater levels? Would rivers become

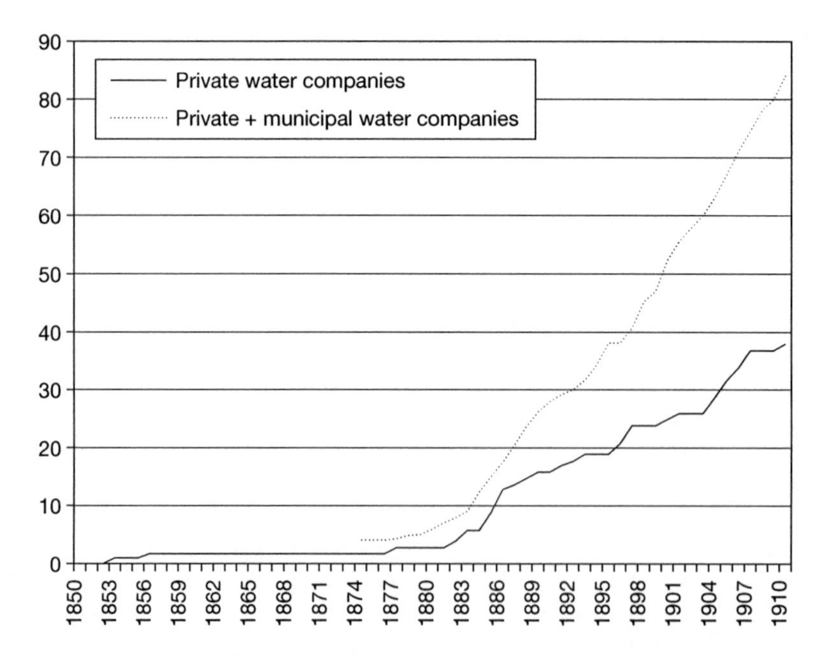

Figure 2.3 Number of private and municipal water companies in the Netherlands.

more contaminated in the future? On various specific dimensions, water supply systems thus differed between various cities.

Although the Dutch transition to piped water systems was carried by various local initiatives, more aggregate national elements also played a role. Quality norms and technical knowledge, for instance, gradually acquired trans-local characteristics. Early definitions of water quality were qualitative, involving turbidity, colour, visible contamination and taste. In the 1880s, engineers proposed new definitions based on chemical properties (e.g. levels of calcium, iron and lead). In the 1890s, microbiology formed a new body of knowledge for definitions of water quality. The initial norm for bacteriological purity proposed in 1895 (100 bacilli per cubic centimetre) was rather arbitrary. In 1904, the Dutch Congress for Public Health published a better-argued Codex, providing indications for chemical and microbiological norms for drinking water (Daru 1992). The creation of the Dutch water association (VWN) in 1899 coincided with the emergence of a technical regime of water engineers and specialists. VWN disseminated knowledge to its members, organized meetings and exchanged experiences. In 1909, VWN developed water quality norms that were indicative, not prescriptive. Formal norms were not articulated until 1957, when a national Piped Water Law formally stated the chemical and biological requirements of drinking water.

The national government only slowly became involved in water-supply matters. It was not until 1901 that it created some pressure through the Housing Law (1901), which articulated quantitative norms concerning the presence of safe drinking water on housing premises and the maximum distance to a source of good water. Although these national regulations provided a basic framework, it was left to city governments to implement them in practice. In 1910, the government created a State Commission for the Question of the Supply of Drinking Water (Wijmer 1992). In 1913, this Commission was split into two bodies: the Central Commission Water Supply, which functioned as an advisory committee to policy makers; and a State Office for Water Supply, which became a technical consultancy agency for urban engineers. The State Office performed technical and scientific research, and set up a central chemical-bacteriological Water Laboratory in 1913 to analyse water samples. The State Office also turned into a product champion, advocating the spread of piped water to rural areas through propaganda campaigns, films and leaflets. But it was not easy to convince farmers that piped water was better than traditional water sources. Water companies were also hesitant, because the connection of rural residents required long pipe infrastructures and high costs. To overcome these financial barriers, the government provided financial subsidies to local water companies in rural areas (Geels 2005). These government activities stimulated the diffusion phase of the transition. The percentage of the Dutch population connected to piped water systems increased from 40 per cent in 1900 to 82.4 per cent in 1951 (Vogelzang 1956).

This diffusion was also stimulated by a new culture of hygiene and cleanliness, which was actively disseminated by doctors, physicians, reformers, churches, housewives' organizations, and so on. A hygienic ideology emerged as middle-class norms of cleanliness linked up with medical science. Cleanliness was important not only for health and hygiene, but also for social distinction (Wright

1984). Proper, healthy and clean behaviour was advocated in a stream of brochures and journals. These hygiene and public health campaigns became part of a broader civilization offensive, which included both hygienic and social disciplining of the urban masses (De Leeuw 1988). The diffusion of piped water thus coincided with a new 'water culture' and user practices centred around water closets, baths, showers, washbasins and washing machines (Meulders 1992). More people began to bathe and wash at home, and markets for new hygienic products (e.g. soap, shampoo, synthetic detergents, perfume, toothbrushes and toothpaste) expanded.

Although cities were a crucial actor in the transition to piped water, the above discussion shows that several national elements and processes (around technical knowledge, policy and culture) were also important.

Cities as initial seedbeds for transitions

Cities are likely to be the initial seedbeds in transitions that start in cities but subsequently gather pace through the involvement of national actors and the creation of ever-larger infrastructures, resulting in national systems. In these transitions, cities act as initial seedbeds for the creation of niches and the performance of entrepreneurial experiments with radically new technologies. Firms and entrepreneurs may see cities as interesting locations because of their concentrated markets and wealth, possible support from policy makers (to solve particular urban problems), and relatively limited infrastructure demands (compared to rural areas). In this pattern, cities are important actors during early phases of these transitions, but diminish in importance in the phase of expansion and up-scaling, which require major investments, and therefore the involvement of national players (government, large industries).

An historical example of this pattern is the transition towards electricity systems between 1880 and 1950. Although the phenomenon of electricity had been known for some time, diffusion proceeded slowly through an accumulation of niche applications. Electricity first played a role in sending messages, via telegraphy in the 1830s and telephony from 1876 on. The next niche was formed by arc lights (in the 1870s), which were used to illuminate important buildings and to create excitement at important festivities. But arc lights were expensive and cumbersome, because users had to generate their own power (using generators or batteries), which required a lot of expertise (Hughes 1983). Arc light therefore remained restricted to incidental demonstration projects.

Following the development of incandescent bulbs in 1878, Edison set out to create the first electric *system* where power was generated by producers, distributed via a grid to users who could power various appliances (initially light). Edison opened the first integrated electricity system in 1882 in New York (the Pearl Street station). This was a local system with a small grid that connected only a few users. The initial niche market attracted interest from shop owners (who used electric light for advertising in shop windows), rich customers who lived in the vicinity and were willing to pay for electric light, from factories for which inflammable gas light posed dangers (e.g. textile factories), and public spaces such as theatres and cafés. City governments subsequently became interested in street lighting, which improved visibility and feelings of safety at night. City governments also played a role in issuing permits for the creation of electricity grids (initially via overhead wires). But the main drivers of this transition were engineers, firms and various user groups.

Electricity systems also spread geographically and popped up in various cities around the world (see Hughes 1983, who analyses in detail the creation and design of electricity systems in Berlin, Chicago and London). Electricity systems were gradually expanded within cities, linking up more neighbourhoods.

In the 1890s, electric trams formed the next important niche, which was also situated within cities, enabling the substitution of horse-based transport (omnibuses and horse trams). Electric trams diffused rapidly in American cities. In 1890, 16 per cent of American street railways were electrified, about 70 per cent were horse- or mule-powered, and 14 per cent consisted of cable cars or steam railways. By 1902, 97 per cent of American street railways were electric (Hilton 1969). Diffusion was rapid, because various powerful groups supported it. Horse-tram companies were eager to replace their expensive horses with electric trams, which had cheaper operating costs. Real estate promoters invested in tramlines because they increased the value of their land. Electric light companies promoted them because they provided an additional market that complemented their night-time light market. Local authorities and urban reformers favoured trams as a means to enhance suburbanization, which they saw as an answer to overcrowded cities.

The next important niche was the use of electricity to power machines, with the introduction of electric motors into factories. During the 1890s, American industries such as printing, publishing and clothing began using electric motors to power machine tools (Geels 2006). These industries were willing to accept the teething problems of early electric motors because they appreciated particular operating characteristics (cleanliness, steady power and speed, ease of control). Between 1899 and 1909, the relative share of electric power in aggregate American manufacturing rose from 5 per cent to 25 per cent (ibid.).

During the first decades of the twentieth century, electricity companies became more interested in economies of scale as a way to rationalize and drive down costs. Takeovers and mergers therefore stimulated the combination of electricity systems on larger scales. This initially occurred within cities, then at the regional scale, and (after World War II) at the national scale. The creation of huge capital-intensive electricity companies coincided with the increasing involvement of industry associations and national governments (Granovetter and McGuire 1998) with regulations and subsidies that facilitated electrification of rural areas. While cities had been important as *locations* for the early phases in the transition, their role was increasingly overtaken by that of these national-level actors.

Limited role for cities

Cities do not play a very important role in transitions that involve limited infrastructure change, that have bigger roles for market dynamics (supply and demand) and that involve transformations of *existing* national-level systems where powerful actors are already established (e.g. incumbent firms, consumer groups).

Cities were not important actors, for instance, in the Dutch transition in production and consumption of meat. Total meat consumption increased from 43 kilograms per year in 1930 to 84 in 1990 (Figure 2.4), with the bulk of this growth situated in the pork segment (from 26 to 45 kilograms per year). On the production side, the number of pigs produced by Dutch farmers increased from 2 million in 1930 to 14 million in 1990, making the Netherlands the biggest European net exporter of pork.

The demand-side changes were related to growth in disposable income, lower relative meat prices due to mass production, marketing strategies from supermarkets (which priced meat low to lure consumers into their shops), cultural changes in the role of family meals and the increased importance of meat in meals (as advocated in cookery books, magazine recipes and culinary advertisements). The supply-side changes entailed a transition from mixed farming to intensive animal husbandry, based on many technical changes such as breeding research, artificial insemination, population genetics, antibiotics, artificial designer foods, indoor husbandry systems, automatic water supply and feeding systems, electric lighting, air conditioning, artificial heaters, and manure removal systems (Geels 2009). Specialization, mechanization, modernization and scale increases characterized this transition, during which the number of farms with pigs decreased by about 80 per cent, from 146,000 in 1960 to 29,000 in 1990. Economic dynamics were important drivers because farmers who adopted the new technologies and increased their operational size produced cheaper pigs and out-competed farmers who did not modernize. But the national government greatly influenced this

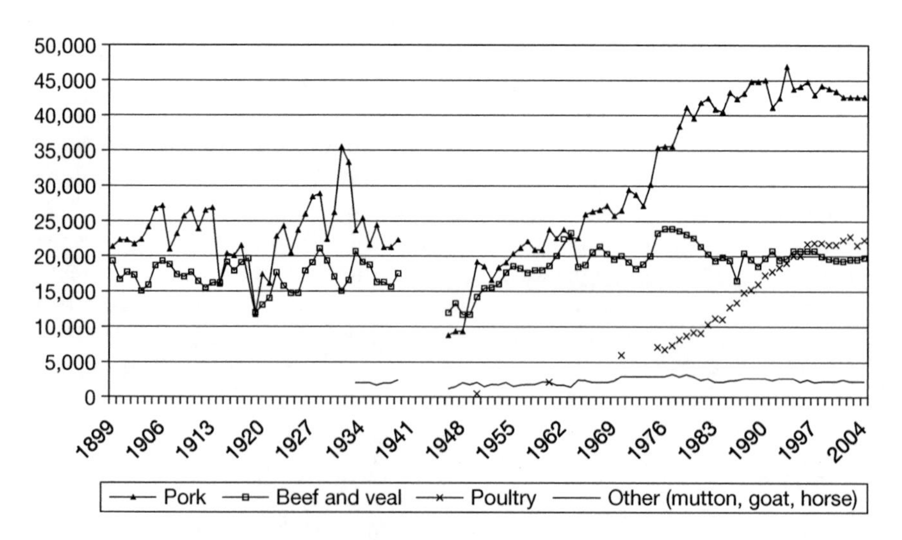

Figure 2.4 Dutch meat consumption per capita per year (in grams).

transition by creating favourable economic frame conditions, including investment subsidies, the tailoring of minimum prices to the economic viability of large modern firms, the provision of guarantees for bank loans, the funding of demonstration projects that showed off new technologies on model farms, a development and buy-out fund that provided subsidies to farmers who wanted to expand, and compensation for farmers who wanted to discontinue operation. These modernization policies were discussed and negotiated in 'iron triangles' that involved national farmers' associations, the Ministry of Agriculture and members of Parliament responsible for agriculture. Institutional power was also used to sideline protesters in the 1970s, including animal welfare groups, environmental groups and the Environmental Ministry (which became concerned about water and air pollution resulting from manure surpluses). The Ministry of Agriculture also tripled investments in research and extension services to boost the development and dissemination of new knowledge and technologies. Extension specialists gave presentations for farmers, visited study clubs, distributed reports and organized excursions to model farms to influence farmers' attitudes towards modernization, mechanization and scale increase. Forward and backward integration also changed the social networks as farmers developed new relations with supermarkets, slaughterhouses, meat-processing companies, technology suppliers, pharmaceutical industry, feed companies, the chemical industry and banks. These changes transformed pig farming into 'agri-business' (Geels 2009). In this transition, which changed technologies, networks, farming practices and consumption patterns, cities played only a limited role.

Conclusions

This chapter has demonstrated that cities can play three roles in technological transitions at the national scale. First, cities can be *primary actors* that enact the transition (especially in cases of regimes characterized by many local systems of provision). Second, cities can be seedbeds and *locations* for radical innovations in the early phases of transitions (but less important during later phases). Third, cities play only limited roles in transitions that depend less on infrastructure, involve strong market interactions, and are about transformations of *existing* systems with powerful incumbent actors.

While book editors claim that cities are important for low carbon transitions because most of the world's population happens to live in cities, this chapter's analytical distinction suggests that this claim needs to be nuanced and differentiated. Instead of a broad-brush claim, the implication of this chapter is that cities can play at least three possible roles with regard to low carbon transitions.

The first role is that city governments and their agencies are important *actors*, something that may be especially relevant with regard to local transport, waste and water systems. Cities can, for instance, attempt to stimulate public transport by improving bus or light-rail systems, or facilitate slow-transport modes such as cycling by creating separate cycle lanes and bicycle parking facilities. City authorities could also decide to stimulate green waste handling or recycling schemes. Some cities also develop visions about energy independence (e.g. London generating its own 'green' electricity). The development of such local energy systems would go against powerful regime tendencies in the past fifty years (such as centralization of electricity grids) and constitute a major transition towards decentralized energy systems. My impression is, however, that there is more talk about cities generating their own energy than concrete action.

The second possibility is for cities to act as early seedbeds for experimenting with and learning about low carbon innovations, providing the *location* where firms, user groups, special-interest groups and local and national governments interact. Cities could, for example, form testing grounds for battery-electric vehicles or fuel-cell vehicles. This could take the form of public tests (such as the European programme for fuel-cell buses in various cities) or private initiatives where motivated consumers would buy or lease battery-electric cars, firms would monitor the user experiences, and employer organizations (or local governments) would provide battery recharging facilities (as in the San Francisco Bay Area). Broader national diffusion of such transport technologies would require the creation of national infrastructures and the involvement of national-level actors (governments, industries, consumer organizations).

The third option is that cities play only limited roles compared to market dynamics and other important actors. This would, for instance, be the case in further expansion of nuclear energy or implementation of carbon sequestration and storage (although cities could be relevant with regard to local planning and permit procedures).

Three further lessons can be derived from the case studies discussed in this chapter. First, transitions are rarely guided by a single goal or purpose. Even in the transition towards piped water, which is often presented as a goal-oriented response to infectious disease problems, cities and various user groups had multiple motivations, of which hygiene was not the most important. This case, but also the discussion of the electric tram, showed that transitions accelerate when actors with different motivations and interests support an innovation. With regard to low carbon transitions, this means that climate change alone may not be sufficient motivation. Low carbon transitions should also be interesting to various groups for other reasons.

Second, the pattern of niche accumulation tends to be important in transitions, as illustrated by the second case. The application of electricity began in small niches but subsequently moved to other niches as the technology improved, as actors gained experience and learned about its potential, and as entrepreneurs saw new potentials. When new lighting systems sprang up in the early 1880s, few people would have imagined how electricity would transform so many domains (e.g. homes, factories, transport, communications). Also, low carbon transitions are likely to be non-linear, to be full of surprises (and setbacks), and to unfold differently from the way experts currently predict.

Third, also in the case where cities were the primary actor (the transition to piped water systems), national elements (in particular, technical knowledge, policy and culture) were important in the diffusion phase. Although the focus on cities and low carbon transitions is important and interesting, this finding suggests that an *exclusive* focus on cities may be unwarranted.

Note

1 The notion of 'levels' in the MLP, which refers to degrees of stability in socio-technical configurations, is different from the notion of geographical scales.

References

Cillekes, C., Van den Boogard, J. and Gales, B. P. A. (1988) *Loop naar de Pomp: Geschiedenis van de Watervoorziening en de Waterleiding in Maastricht* (Pumping water: A history of water provision and waterworks in Maastricht), Maastricht, the Netherlands: Stichting Historische Reeks.

Daru, M. (1992) 'Risky business: Piped drinking water in Dutch towns, 1880–1920', Paper presented at the First Conference of the European Association of Urban Historians, Amsterdam, September.

De Leeuw, K. (1988) 'Hygiene en gezondheid als terrein van beschavingsoffensief: Regulering en disciplinering in Nederland en Noord-Brabant 1880–1940' (Hygiene and health as domains of the civilization offensive: Regulation and discipline in the Netherlands and North Brabant 1880–1940), *Sociale Wetenschappen* 31 (3): 145–152.

Elzen, B., Geels, F. W. and Green, K. (eds) (2004) *System Innovation and the Transition to Sustainability: Theory, Evidence and Policy*, Cheltenham, UK: Edward Elgar.

Geels, F. W. (2002) 'Technological transitions as evolutionary reconfiguration processes: A multi-level perspective and a case-study', *Research Policy* 31 (8–9): 1257–1274.

Geels, F. W. (2004) 'From sectoral systems of innovation to socio-technical systems: Insights about dynamics and change from sociology and institutional theory', *Research Policy* 33 (6–7): 897–920.

Geels, F. W. (2005) 'Co-evolution of technology and society: The transition in water supply and personal hygiene in the Netherlands (1850–1930) – a case study in multi-level perspective', *Technology in Society* 27 (3): 363–397.

Geels, F. W. (2006) 'Major system change through stepwise reconfiguration: A multi-level analysis of the transformation of American factory production (1850–1930)', *Technology in Society* 28 (4): 445–476.

Geels, F. W. (2009) 'Foundational ontologies and multi-paradigm analysis, applied to the socio-technical transition from mixed farming to intensive pig husbandry (1930–1980)', *Technology Analysis & Strategic Management* 21 (7): 805–832.

Geels, F. W. and Raven, R. P. J. M. (2006) 'Non-linearity and expectations in niche-development trajectories: Ups and downs in Dutch biogas development (1973–2003)', *Technology Analysis & Strategic Management* 18 (3–4): 375–392.

Geels, F. W. and Schot, J. W. (2007) 'Typology of sociotechnical transition pathways', *Research Policy* 36 (3): 399–417.

Geels, F. W., Hekkert, M. and Jacobsson, S. (2008) 'The micro-dynamics of sustainable innovation journeys: Editorial', *Technology Analysis & Strategic Management* 20 (5): 521–536.

Granovetter, M. and McGuire, P. (1998) 'The making of an industry: Electricity in the United States', in M. Callon (ed.) *The Laws of the Markets*, Oxford: Blackwell.

Groen, J. A. (1980) *Een Cent per Emmer: Het Amsterdamse Drinkwater door de Eeuwen Heen* (A cent per bucket: Drinking water in Amsterdam through the Centuries), Amsterdam: Gemeentewaterleidingen.

Groote, P. D. (1995) *Kapitaalvorming in Infrastructuur in Nederland, 1800–1913* (Capital investment in infrastructure in the Netherlands, 1800–1913), Capelle aan den IJssel: Labyrint.

Hilton, G. W. (1969) 'Transport technology and the urban pattern', *Journal of Contemporary History* 4: 123–135.

Hughes, T. P. (1983) *Networks of Power: Electrification in Western Society, 1880–1930*, Baltimore: Johns Hopkins University Press.

Kemp, R. (1994) 'Technology and the transition to environmental sustainability: The problem of technological regime shifts', *Futures* 26 (10): 1023–1046.

Meulders, C. (1992) *The Struggle for Cleanliness: A Socio-historical Analysis of the Laundry Process*, Leuven, Belgium: Catholic University of Leuven.

Rip, A. and Kemp, R. (1998) 'Technological change', in S. Rayner and E. L. Malone (eds) *Human Choice and Climate Change*, vol. 2, Columbus, OH: Battelle Press.

Shackley, S. and Green, K. (2007) 'A conceptual framework for exploring transitions to decarbonised energy systems in the United Kingdom', *Energy* 32 (3): 221–236.

Smith, A., Stirling, A. and Berkhout, F. (2005) 'The governance of sustainable socio-technical transitions', *Research Policy* 34 (10): 1491–1510.

Tarr, J. A. (1981) 'Changing fuel use behavior: The Pittsburgh smoke control movement, 1940–1950', *Technological Forecasting and Social Change* 20 (4): 331–346.

Van den Noort, J. (1990) *Pion of pionier: Rotterdam-Gemeentelijke Bedrijvigheid in de Negentiende Eeuw* (Pawn or pioneer: Rotterdam's urban activities in the nineteenth century), Rotterdam: Stichting PK.

Van der Heijden, C. G. W. P. (1995) *Kleurloos, reukloos en smaakloos drinkwater: De watervoorziening in Tilburg vanaf het weinde van de Negentiende eeuw* (Colourless, odourless and tasteless drinking water: Drinking water provision in Tilburg since the late nineteenth century), Tilburg: Stichting tot Behoud van Tilburg Cultuurgoed.

Van Driel, H. and Schot, J. (2005) 'Radical innovation as a multi-level process: Introducing floating grain elevators in the port of Rotterdam', *Technology and Culture* 46 (1): 51–76.

Verbong, G. P. J. and Geels, F. W. (2007) 'The ongoing energy transition: Lessons from a socio-technical, multi-level analysis of the Dutch electricity system (1960–2004)', *Energy Policy* 35 (2): 1025–1037.

Vogelzang, I. (1956) *De Drinkwatervoorziening van Nederland voor de Aanleg van de Drinkwaterleidingen* (Drinking water systems in the Netherlands before the construction of piped water systems), Utrecht: Rijksuniversiteit Utrecht.

Wijmer, S. (1992) *Water om te Drinken* (Water for drinking), Rijswijk: Vereniging van Exploitanten van Waterleidingbedrijven in Nederland (VEWIN).

Wiskerke, J. S. C. and Van der Ploeg, J. D. (2004) *Seeds of Transition: Essays on Novelty Production, Niches and Regimes in Agriculture*, Assen, the Netherlands: Van Gorcum.

Wright, L. (1984) *Clean and Decent: The History of the Bathroom and the WC*, London: Routledge & Kegan Paul.

York, R., Van Driel, H. and Rosa, E. A. (2003) 'Key challenges to ecological modernization theory', *Organization & Environment* 16 (3): 273–288.

3 Governing urban low carbon transitions

Harriet Bulkeley, Vanesa Castán Broto and Anne Maassen

Introduction

Cities are increasingly being recognised as critical arenas for addressing climate change. Since 2008, for the first time half of the world's population is living in towns and cities (UN 2008). The concentration of social and economical activities in cities has led to the recognition of urban areas as key sites in the production of greenhouse gas emissions and as potentially vulnerable to the impacts of climate change (e.g. Stern 2006; IEA 2008). In this context, there is a growing acknowledgement of the opportunities that cities provide for addressing climate change. Cities are portrayed as financial centres and centres of technical, social and political innovation (OECD 2008), and they are regarded as providing 'hubs' where innovative approaches to climate change mitigation can be 'tested' (e.g. DECC 2009) and adaptation measures adopted. Furthermore, city governments have had an important role in creating opportunities for climate change action (e.g. Bulkeley and Betsill 2003). The emergence of global city networks such as ICLEI Local Governments for Sustainability, the Climate Alliance and the C40 has demonstrated the importance of cities in driving global responses to climate change, sometimes filling the vacuum left by national governments' inaction.

Mirroring these debates, there is growing scholarship on cities and climate change which focuses on urban responses to climate change and the factors that shape urban governance of climate change (e.g. Bulkeley and Betsill 2003; Bai 2007; Betsill and Bulkeley 2007; Granberg and Elander 2007; Holgate 2007; Romero Lankao 2007; Schreurs 2008; Bai *et al.* 2009). This work suggests that urban climate change responses can be understood primarily in terms of the development of new forms of policy and planning, and thus processes of change are driven primarily by institutional and political processes. Within this approach, low carbon transitions have been regarded fundamentally in terms of 'policy change', with consideration being given to the different factors (e.g. leadership, capacity, resources) that facilitate or hinder such changes (Bulkeley *et al.* 2009). However, the dynamics of urban infrastructure and the challenges emerging from conflicts over the political economy of the city have not been fully explored (Bulkeley forthcoming; Monstadt 2009). In this chapter, we argue that alternative accounts of 'transitions' and of the city can provide a means through which to engage with the political and material basis of urban responses to climate change: first, insights from debates concerning innovation and transition in socio-technical systems; and second, an engagement with the ways in which socio-environmental systems are transformed in the city. Drawing on these perspectives and focusing on urban infrastructure systems as critical in mediating the production of greenhouse gas emissions, we suggest that efforts to mitigate climate change in the city are shaped by the dynamic tensions between processes of experimentation and efforts to engender systemic change towards low carbon futures; by the stability, or obduracy, of these networks; by political economies of urban restructuring; and by the constant flux of 'metabolic circulation' in the city. Accomplishing urban low carbon transitions becomes a matter not only of policy, or of 'niche' experimentation, but of the reconfiguration of socio-technical networks – a process that is at once highly political and open to contestation and disruption.

In the next section of this chapter, we begin to develop this argument further by examining the potential of theories of socio-technical transition in the context of urban responses to climate change. We argue that such approaches provide a useful emphasis on the technical and material components of system transition, but to date have been limited by their conception of the urban, and by limited analyses of the political dynamics of niche and regime transformation. Attending to these limitations, we suggest, requires that we examine in more detail the nature of urban socio-technical systems and their dynamics. To this end, the third section explores the processes whereby urban infrastructures reconfigure the city, and argues that managerial approaches to urban transitions are challenged by both nature's agency and everyday practices of cities' inhabitants. Finally, the fourth section discusses the implications of these debates for developing our understanding of low carbon socio-technical transitions and their consequences.

From policy change to socio-technical change

Urban climate governance studies have emerged in parallel with a growing interest in cities as arenas in which to manage climate change mitigation and adaptation. These studies have focused on assessing the impact of urban policies and initiatives (for a review, see Betsill and Bulkeley 2007; Bulkeley forthcoming), on the integration of these systems of governance across scales, for example in city networks (e.g. Bäckstrand 2008; Andonova *et al.* 2009; Kern and Bulkeley 2009), or on reviewing the different factors that facilitate or halt climate change governance within the city (e.g. Bulkeley and Betsill 2003; Romero Lankao 2007; Rutland and Aylett 2008). These examples have shown that political and social innovation lie at the heart of these climate change initiatives. Hoffman (forthcoming) has argued that we are entering a period of climate governance *experimentation* taking place at all levels of political organisation.[1] At the municipal level, climate governance experiments take various forms, such as new forms of political organisation, the creation and use of new governance instruments operating both across and within cities, and local and regional public–private initiatives (ibid.). Some examples of such experiments include the use of the municipal planning system to enforce higher energy efficiency standards and the use of low carbon energy sources, traffic management strategies such as congestion charging, and the creation of new forms of public–private partnerships. An example of the latter is a Woking (Surrey, UK) energy services company (ESCO) that has pioneered the delivery of decentralised energy solutions in the United Kingdom in the shape of a public–private venture which invests in energy and environmental service projects, such as district heating and cooling plants and networks.

As these experiments indicate, governing climate change in the city not only is a matter of institutional or political change, but also involves processes of social and technological innovation and transformation. On this basis, we suggest that a socio-technical understanding of the urban can provide a useful means through which to examine the political and material processes associated with climate change-related innovation. Analyses of socio-technical systems, such as (urban) infrastructure networks, owe much to the pioneering work of Hughes (1983, 1987), who described Large Technical Systems as clusters of artefacts and social, cultural, economic and political factors that are both socially constructed and society shaping. Large Technical Systems are relatively stable, or obdurate, systems where change is largely gradual and driven by external factors. Nonetheless, such systems, including the provision of water and energy, or transportation, have been transformed over time. Recent studies have developed a 'multilevel perspective' through which to understand such transitions (see Geels, this volume, Chapter 2). This perspective describes socio-technical systems as structured through an enduring landscape of 'external' factors, organised in 'socio-technical regimes' characterised by a series of semi-consistent social rules (shared values, common problem definitions, standards, knowledge and skills) that give them coherence and stability, and punctuated by 'niches' within which experimentation takes place (Geels 2004, 2005; see also Geels, this volume, Chapter 2). Like Large Technical Systems, socio-technical regimes tend to both physical and social inertia, leading to 'path dependency' (Berkhout 2002) and 'lock-in' (Unruh 2002). However, this stability is not a given, but rather is the result of the constant 'work' of the regime in maintaining its dominance. At the same time, the regime is open to continuous disturbance, which most often leads to 'intra-systemic' adaptation and transformations in incremental steps, in which the regime appears to remain relatively stable (Geels and Kemp 2007). However, the combined effect of actors' actions in niches and the aperture of 'windows of opportunity' in the regime and/or the landscape may cause a rapid and abrupt change leading to the realignment of the three levels in a new and stable socio-technical regime configuration (Geels 2002, 2005; Hoogma *et al.* 2002; Geels and Schot 2007).

This is often demonstrated by analysing the historical trajectories of specific technologies (see Geels, this volume, Chapter 2). However, more recently authors have argued that niches can be managed to support transitions through a process of 'strategic niche management' (Raven 2007). One area in which this approach has been advanced is in relation to sustainable energy systems (Schot 1998; Hegger *et al.* 2007; Raven 2007). This approach promotes 'top-down' interventions to create or manage these niches (through, for example, government grants), prioritising technological experimentation over socio-technical change. Other scholars have focused on social niches emerging on the margins of the mainstream socio-technical configuration, where citizens and NGOs experiment with environmental technology outside conventional business and government institutions' arrangements (Hegger *et al.* 2007). Here, the focus is on social, rather than technological, experimentation. For example, 'grassroots innovations' are community-led initiatives with the potential for wider transformation of mainstream society such as straw-bale housing or local renewable energy projects (Seyfang and Smith 2007). In other cases, third-sector organisations, individual pioneers or business interests may seek to develop 'intermediate projects' to bring niche-generated principles and framings into the dominant regime (Smith 2007). The combined experience of strategic niche management and social niches suggests that both the provision of some form of protection or shelter from the wider socio-technical system (conceived in the form of government subsidies and preferential treatment) and the internal process of learning by actors gathered around the emerging (socio-) technology are critical to foster spaces of socio-technical experimentation (Hegger *et al.* 2007; Hodson and Marvin 2009).

Research on social niches and strategic niche management suggests that transitions can be deliberate, because the alignment of actors and artefacts into a new regime involves 'many mutual translations between actors' (Smith 2007: 448). Translation becomes a strategic practice through which niche practices can be interpreted, adapted and accommodated within the socio-technical regime. This is certainly the approach taken by the UK government in the 2009 *Low Carbon Transition Plan*, which in a managerial style delineates a 'route-map' for a UK low carbon transition between now and 2020. The plan, however, is not restricted to measures over which the government has direct control (such as financing mechanisms for green business or house retrofitting) but also relies on the idea that individuals, communities and business will follow suit (DECC 2009).

The strength of the multilevel model is its capacity to describe the processes of change and stability within the system and its ability to identify niches and processes of experimentation as routes for deliberate interventions to spur abrupt change. However, there is a question about the extent to which niche and regime can be neatly separated. While both niches and regimes consist of the same 'building blocks' (actors, technical artefacts, material constraints, social practices and institutions, technology interpretations), the distinguishing feature seems to be their relative stability (Geels 2004, 2005). Innovations build on existing ideas, values and technologies, so few of them are truly radical (Lovell 2009): niches emerge from existing socio-technical regimes. If, in practice, it is difficult to determine the boundaries of niches and their relation to socio-technical regimes, their role in contributing to broad systemic change is once again open to question. At the same time, the key role ascribed to government actors in creating 'protected' spaces for niche development raises questions as to whether niches are established in order to maintain regimes rather than as a means of fostering change. Debates about socio-technical systems therefore need to engage with who has agency to lead or manage the process of innovation and catalyse specific types of transition.

Moreover, while the obduracy of socio-technical systems is recognised, this is primarily related to the ways in which particular (technical) innovations 'take' within existing markets or institutional settings. In conceiving of systems as both dynamic and obdurate, the model underplays the importance of conflict as an underlying motivation and structural constraint of change. For example, 'green niches', most notably 'grassroots innovations', are often constructed in opposition to the mainstream regime, as responses to perceived problems within the regime (Smith 2007). Indeed, 'upscaling' niche experiments to trigger a regime transition is sometimes perceived as a threat to the innovative and local transformative potential of niche experiments (Seyfang 2009). Innovation is also a political process in which varied actors with contrasting perspectives intervene. Thus, there is a need to understand the political context of system innovation and, in particular, how contestation practices shape the process of innovation in unexpected ways. Furthermore, although there is recognition of the context-based and local character of some transitions (Raven 2007; Raven *et al.* 2008), the literature on systems innovation has not formally examined the spatial geographies of transition (Hodson and Marvin, this volume, Chapter 5; Smith *et al.* 2010). With its implicit national focus, the multilevel perspective regards urban energy systems as relatively indistinguishable from the national system of which they are a part. The way in which urban configurations of actors, artefacts, materials, and so on may serve to produce distinct 'urban socio-technical regimes' (Monstadt 2009) is not recognised. Cities are approached either as homogeneous actors that act with a certain degree of autonomy in influencing government choices, or as the space of specific types of innovation (see Geels, this volume, Chapter 2). In seeking to understand specific urban responses to climate change, we argue, an alternative account of the city is needed, one that can also take into account the politics of experimentation and of obduracy.

Configuring the city: nature, infrastructure and the urban

The literature on climate change and environmental governance has demonstrated the relevance of the spatial component, the local and the urban, in transitions to low carbon systems. However, to date, its conception of the urban has been limited to one that delimits the city in terms of a polity or space for governing, neglecting its socio-technical dimensions. As we argued earlier in the chapter, insights from the analysis of the transformation of (large) socio-technical systems can provide a material counterbalance to these analyses, but they are also limited in the ways in which cities, and the urban infrastructure systems of which they are composed, are conceived. Rather than being fundamentally socio-technical, cities are either regarded in terms of their administrative boundaries as sites for innovation, or considered as one (set of) political actors. Here, we suggest that in order to understand urban low carbon transitions we need instead to conceptualise the city and infrastructure systems as co-constitutive, reinstating the material and spatially fixed components of the lived experiences of the city (Gandy 2004), and recognising the essentially contested nature of these processes.

One avenue through which to examine urban transitions, the politics of urban infrastructure and how this is materialised in socio-technical systems, lies in the process that Graham and Marvin (2001) have called splintering urbanism. According to this approach, the 'modern infrastructure ideal', in which ideas of universal and equal provision of services directed state action on urban infrastructures, has given way, heralding a transition in urban infrastructure systems. Taken-for-granted discourses of profitability are leading to a widespread movement towards privatisation and liberalisation which results in the unbundling of infrastructure networks and the fragmentation of the urban (ibid.). Splintering urbanism is a dual process in which existing integrated infrastructure networks are unbundled and segmented (institutionally by delegating states' responsibility to contractors and corporations, or materially by creating physical or virtual bypasses), while simultaneously there is a 'rebundling' of the city through the creation of 'premium networked spaces' (such as gated communities, resorts or theme parks) that actively separate the economic lives of the rich from those of the poor (ibid.). This process leads to the mapping of urban centres into differential spaces of affluence and marginality which, combined with the growing influence of surveillance and control technologies, 'support the sociotechnical partitioning of the metropolitan, and indeed, societal, fabric' (ibid.: 383).

However, the applicability of these concepts across different regions has been questioned. Empirical explorations of urbanism in cities in the global South have argued that this analysis cannot accurately reflect the socio-historical context in which the infrastructure has (or rather, has not) developed in these cities. Often, colonial state ideas were informed by the purification of social groups into coloniser and colonised categories which helped to develop mechanisms that excluded the latter from the systems of 'universal' provision (McFarlane and Rutherford 2008; Zérah 2008). Moreover, cities in the global South may have never been integrated in a national-base networked system for a variety of reasons, including rapid population growth, lack of resources, poor political will, fiscal constraints or even the complete absence of the 'modern infrastructural ideal' rhetoric (see Kooy and Bakker 2008; McFarlane and Rutherford 2008; Zérah 2008). Despite these limitations, by drawing attention to the ways in which the development of urban infrastructure systems is structured through wider political and economic processes, and is taking different forms in different urban regions, the splintering urbanism approach raises important questions in terms of how we might understand the processes of low carbon urban transitions. For example, recognising the differences in urban contexts, and in particular the lack of infrastructure and services in the global South, questions the emphasis in climate change narratives on adapting infrastructure to climate change or implementing mitigation initiatives in infrastructure that has never been there (Satterthwaite 2008). Although there is potential for low carbon innovations to favour groups of population whose access to energy provision is restricted (take, for instance, the low-cost installation of solar water heaters in deprived neighbourhoods, which could help palliate energy poverty problems), the current 'crisis of infrastructure' (Guy *et al.* 1997; see also Graham and Marvin 2001) impedes, rather than being solved by, current climate change policies.

A further set of considerations arises if we consider renewable energy sources as part of potential low carbon energy futures. Often, the inability of these technologies to 'break through' the current regime has been attributed to institutional barriers, in line with notions of 'implementation gaps' in policy analyses and the barriers and drivers to innovation identified by the multilevel perspective. However, the growing emphasis on the need to develop a new 'low carbon' energy paradigm that relies largely on the decentralisation of the systems of provision and the liberalisation of service providers could also be regarded as part of a process of splintering (urban) networks (Table 3.1). In the case of the United Kingdom, Elliott (2000) argues that although decentralisation has already occurred as the result of the combined action of the implementation of liberalisation narratives and the shift from coal to gas, the introduction of renewable energy could facilitate further decentralisation (and fragmentation) of energy systems. This narrative certainly resonates with the Greater London Authority plans for decentralised energy (London Development Agency 2010). In this context, climate change experiments may perpetuate current trends in market liberalisation and the idea that this, by itself, may lead to lesser (and local) environmental impacts (see Table 3.1). The extent to which low-carbon and decentralised energy systems are being developed in the urban arena may therefore have more to do with broader processes of economic and political restructuring than with the dynamics of 'niche' innovation.

In the urban context, analysis suggests that these processes are manifest through 'strategic relocalisation', on the one hand, whereby infrastructure is rescaled 'upwards' to create systemic transition from local experiments or exemplars and 'downwards' by withdrawing from existing regional and national infrastructure; and 'glurbanisation', on the other hand, fostered by the creation of international networks of selected cities in which cities work collaboratively to share developments of new infrastructure fixes and 'downscale' shared knowledge into other cities within the national urban hierarchies (Hodson and Marvin 2007, 2009). While the former concept resonates with understandings of the role of niches in

Table 3.1 A changing energy paradigm?

	Conventional paradigm	*New paradigm*
Fuel resource	Finite stocks	Renewable flows
Energy type	Concentrated	Diffuse
Technology	Large scale	Smaller scale
Generation	Centralised	Decentralised
Environmental impact	Large, global	Small, local
Market	Monopoly	Liberalised

Source: Adapted from Elliott (2000).

innovation processes within the multilevel model of sustainable transitions, the latter echoes claims advanced within the climate change governance literature which explore the role of cities and city networks in global environmental governance (Bulkeley and Betsill 2003; Bulkeley 2005; Kern and Bulkeley 2009). Importantly, however, splintering urbanism and other accounts of the development of urban infrastructure systems point to the highly political nature of these processes (Guy *et al.* 1997; Moss 2003; McFarlane and Rutherford 2008). Splintering urbanism suggests that selected actors (e.g. coalitions of corporate, political leaders and environmental groups) are often able to shape the dominant responses to the ecological and infrastructure security of cities (Guy *et al.* 1997; Hodson and Marvin 2009). For example, Hodson and Marvin (2007) found that corporate capital interests as technology providers predominated in the negotiations of the development of hydrogen energy infrastructure in London, while local resistance was effectively circumscribed by claims over the national and international importance of this low carbon 'experiment'.

Yet there is a diversity of actors who shape the development of infrastructures and urbanisation, whether this is in explicit contestation actions or in the resistance practices embedded in everyday life in the city: 'Because infrastructure is big, layered and complex, and because it means different things locally, it is never changed from above. Changes take time and negotiation, and adjustments with other aspects of the systems are involved' (Star 1999: 382). In contrast to the splintering urbanism thesis, other work on urban infrastructure has sought to engage with the experiences and subjectivities of different urban actors whose practices and interpretations define the use and development of infrastructure (Giglioli and Swyngedouw 2008; Kooy and Bakker 2008; McFarlane 2008; McFarlane and Rutherford 2008). Some scholars have sought to engage with the operations of non-human actors and their influence in urban life (Gandy 2005; Swyngedouw 2006; McFarlane 2008). Engagement with questions about nature in the city has led to a characterisation of the city as a form of human ecological organisation surrounded by processes that are simultaneously local and global, semiotic and material (Heynen *et al.* 2006). Here, the city is described as a process of socio-ecological change (Swyngedouw and Heynen 2003). These socially mediated processes of environmental and technological transformation are referred to as 'metabolic circulation' (Swyngedouw 2006). The idea of 'circulation' emphasises the centrality of infrastructures to the routine exchange processes in the city, joining the ecological flows that mediate the transformation of nature into city (Kaika and Swyngedouw 2000). However, they also point that these flows are not merely material. Rather, the focus on urban metabolism brings uneven power relationships within the city at the centre of the analysis, but furthers the thesis of splintering urbanism by linking it with ecological processes within the city and a questioning of the boundaries and agency of human and non-human actors.

Rather than focusing on the obdurate, stable nature of urban infrastructure systems, or the ways in which they are shaped by structural economic and political processes, the analysis of urban socio-natures directs attention to the 'uncanny' in the city (Kaika 2006). Infrastructure networks are not only contested but also in some way subverted by the everyday practice of actors in the city. Contestation comes from the actions of social groups and individuals that shape the city through the continuous engagement with its material fabric which (deliberately or not) may challenge the discursive and material practices of those who, top-down or bottom-up, intervene in the governing of transitions to low carbon systems. In this way, accounts of urban metabolism draw attention to how individuals and groups within the city incorporate a range of daily practices that are not necessarily aligned with the visions of sustainable futures promoted from within the mainstream regime. Urban metabolism has to be essentially regarded as a process of nature production. Nature is inscribed in the city as simultaneously desirable and uncontrollable (Desfor and Keil 2004; Kaika 2006). Nature's disruption of urban infrastructure, such as in the famous example of the New York blackout (Graham 2010), highlights the fundamentally unruly nature of socio-technical systems. A managerial approach to transitions resonates with unfulfilled Promethean accounts of nature domination within the urban (Kaika 2006) and contradicts Hughes' (1987) description of socio-technical processes as subject to 'messy' processes. Recognising the experimental and uncertain character of transitions, raising doubt as to whether transitions can be predicted or controlled, may open the door for the recognition of the co-construction of the material and the social in urban infrastructure.

Conclusions: urban energy systems in transition

Our analysis of the process of urban low carbon transitions suggests that we need to move beyond a framework that locates the drivers of change within the institutional structures and governance arrangements through which policies are developed and implemented, to consider the fundamentally socio-technical nature of interventions in urban infrastructure systems. The multilevel perspective on systems in transition provides a critical starting point for such an analysis, demonstrating the ways in which infrastructure systems tend to obduracy but can also undergo dynamic change, by highlighting how the overlapping of social and technological elements shape the dynamics of stability and transition. The literature has provided insights into how a future low carbon transition could be brought about by focusing on the deliberate creation of niches, or protected spaces, directed either by the government or by civil society organisations, in which new low carbon innovations can flourish until a window of opportunity allows the active transformation of the regime. However, we suggest that the multilevel perspective fails to engage sufficiently with the 'urban' nature of these processes, or their fundamentally political character. In order to address these shortcomings, we suggest that insights from the social study of urban infrastructure systems may be useful, particularly in revealing the ways in which obduracy and flux in urban systems is created, maintained and contested. On the one hand, accounts of splintering urbanism point to the ways in which processes of economic and political restructuring are enacted through, and mediated by, urban infrastructure systems. In this reading, attempts to 'experiment' in response to climate change must be contextualised with regard to the broader political economy that such efforts sustain or contest. Developing these perspectives can provide critical assessments of the social and environmental justice of socio-technical transitions (who wins and who loses), and in this manner can contribute to existing debates on the governing of climate change. On the other hand, conceiving of the urban in socio-ecological terms points to the role of everyday practices and of nature as 'the uncanny' in shaping urban materiality. We can therefore conceive of urban low carbon transitions as forged through the interaction of processes of strategic intervention, and of crisis, disruption and contestation. In this context, managerial approaches to transitions to sustainability need to be balanced with the realisation of the fundamentally chaotic nature of transitions processes regarding urban infrastructure. Experimentation, as an open-ended process, becomes central both to urban environmental governance and for the socio-technical and socio-ecological transformations of urban infrastructure.

Note

1 Hoffman's analysis explicitly excludes climate governance experimentation taking place at the level of individual municipal authorities, on the basis, he argues, that this is taking place within existing government structures and through established means of exercising power and authority. However, we seek to extend his argument to the urban arena on the basis that many climate change interventions are conducted outside the formal authority of the (local) state, involve novel governance arrangements across public/private divides and, importantly in the context of this chapter, socio-technical innovation.

References

Andonova, L. B., Betsill, M. M. and Bulkeley, H. (2009) 'Transnational climate governance', *Global Environmental Politics* 9 (2): 52–73.

Bäckstrand, K. (2008) 'Accountability of networked climate governance: The rise of transnational climate partnerships', *Global Environmental Politics* 8 (3): 74–102.

Bai, X. (2007) 'Integrating global environmental concerns into urban management: The scale and readiness arguments', *Journal of Industrial Ecology* 11 (2): 15–29.

Bai, X., Wieczorek, A. J., Kaneko, S., Lisson, S. and Contreras, A. (2009) 'Enabling sustainability transitions in Asia: The importance of vertical and horizontal linkages', *Technological Forecasting and Social Change* 76 (2): 255–266.

Berkhout, F. (2002) 'Technological regimes, path dependency and the environment', *Global Environmental Change* 12 (1): 1–4.

Betsill, M. and Bulkeley, H. (2007) 'Looking back and thinking ahead: A decade of cities and climate change research', *Local Environment: The International Journal of Justice and Sustainability* 12 (5): 447–456.

Bulkeley, H. (2005) 'Reconfiguring environmental governance: Towards a politics of scales and networks', *Political Geography* 24 (8): 875–902.

Bulkeley, H. (forthcoming) 'Cities and the governing of climate change' (under review).

Bulkeley, H. and Betsill, M. M. (2003) *Cities and Climate Change: Urban Sustainability and Global Environmental Governance*, London: Routledge.

Bulkeley, H., Schroeder, H., Janda, K., Zhao, J., Armstrong, A., Chu, S. Y. and Ghosh, S. (2009) 'Cities and climate change: The role of institutions, governance and urban planning', Paper presented at the World Bank Fifth Urban Research Symposium, 'Cities and Climate Change: Responding to an Urgent Agenda', Marseille.

DECC (2009) *The UK Low Carbon Transition Plan: National Strategy for Climate and Energy*, London: Department for Energy and Climate Change.

Desfor, G. and Keil, R. (2004) *Nature and the City: Making Environmental Policy in Toronto and Los Angeles*, Tucson: University of Arizona Press.

Elliott, D. (2000) 'Renewable energy and sustainable futures', *Futures* 32 (3–4): 261–274.

Gandy, M. (2004) 'Rethinking urban metabolism: water, space and the modern city', *City* 8: 363–379.

Gandy, M. (2005) 'Cyborg urbanization: Complexity and monstrosity in the contemporary city', *International Journal of Urban and Regional Research* 29 (1): 26–49.

Geels, F. W. (2002) 'Technological transitions as evolutionary reconfiguration processes: A multi-level perspective and a case-study', *Research Policy* 31 (8–9): 1257–1274.

Geels, F. W. (2004) 'From sectoral systems of innovation to socio-technical systems: Insights about dynamics and change from sociology and institutional theory', *Research Policy* 33 (6–7): 897–920.

Geels, F. (2005) *Technological Transitions and System Innovations: A Co-evolutionary and Socio-technical Analysis*, Cheltenham, UK: Edward Elgar.

Geels, F. W. and Kemp, R. (2007) 'Dynamics in socio-technical systems: Typology of change processes and contrasting case studies', *Technology in Society* 29 (4): 441–455.

Geels, F. W. and Schot, J. (2007) 'Typology of sociotechnical transition pathways', *Research Policy* 36 (3): 399–417.

Giglioli, I. and Swyngedouw, E. (2008) 'Let's drink to the great thirst! Water and the politics of fractured techno-natures in Sicily', *International Journal of Urban and Regional Research* 32 (2): 392–414.

Graham, S. (2010) 'Disruption by design: Urban infrastructure and political violence', in S. Graham (ed.) *Disrupted Cities: When Infrastructure Fails*, New York: Routledge.

Graham, S. and Marvin, S. (2001) *Splintering Urbanism: Networked Infrastructures, Technological Mobilities and the Urban Condition*, London: Routledge.

Granberg, M. and Elander, I. (2007) 'Local governance and climate change: Reflections on the Swedish experience', *Local Environment: The International Journal of Justice and Sustainability* 12 (5): 537–548.

Guy, S., Graham, S. and Marvin, S. (1997) 'Splintering networks: Cities and technical networks in 1990s Britain', *Urban Studies* 34: 191–216.

Hegger, D. L. T., Van Vliet, J. and Van Vliet, B. J. M. (2007) 'Niche management and its contribution to regime change: The case of innovation in sanitation', *Technology Analysis & Strategic Management* 19 (6): 729–746.

Heynen, N., Kaika, M. and Swyngedouw, E. (2006) 'Urban political ecology: Politicizing the production of urban natures', in N. Heynen, M. Kaika and E. Swyngedouw (eds) *In the Nature of Cities: Urban Political Ecology and the Politics of Urban Metabolism*, London: Routledge.

Hodson, M. and Marvin, S. (2007) 'Understanding the role of the national exemplar in constructing "strategic glurbanization"', *International Journal of Urban and Regional Research* 31: 303–325.

Hodson, M. and Marvin, S. (2009) ' "Urban ecological security": A new urban paradigm?', *International Journal of Urban and Regional Research* 33 (1): 193–215.

Hoffman, M. J. (forthcoming) *Climate Governance at the Crossroads: Experimenting with a Global Response*, New York: Oxford University Press.

Holgate, C. (2007) 'Factors and actors in climate change mitigation: A tale of two South African cities', *Local Environment: The International Journal of Justice and Sustainability* 12 (5): 471–484.

Hoogma, R., Kemp, R., Schot, J. and Truffer, B. (2002) *Experimenting for Sustainable Transport: The Approach of Strategic Niche Management*, London: Spon Press.

Hughes, T. (1983) *Networks of Power Electrification in Western Society, 1880–1930*, Baltimore: Johns Hopkins University Press.

Hughes, T. (1987) 'The evolution of large technical systems', in W. Bijker, T. P. Hughes and T. Pinch (eds) *The Social Construction of Technological Systems: New Directions in the Sociology and History of Technology*, Cambridge, MA: MIT Press.

IEA (2008) *World Energy Outlook 2008*, Paris: International Energy Agency.

Kaika, M. (2006) *City of Flows*, London: Routledge.

Kaika, M. and Swyngedouw, E. (2000) 'Fetishizing the modern city: The phantasmagoria of urban technological networks', *International Journal of Urban and Regional Research* 24 (1): 120–138.

Kern, K. and Bulkeley, H. (2009) 'Cities, Europeanization and multi-level governance: Governing climate change through transnational municipal networks', *JCMS: Journal of Common Market Studies* 47 (2): 309–332.

Kooy, M. and Bakker, K. (2008) 'Technologies of government: Constituting subjectivities, spaces, and infrastructures in colonial and contemporary Jakarta', *International Journal of Urban and Regional Research* 32 (2): 375–391.

London Development Agency (2010) 'Decentralised energy'. Online, available at: www.lda.gov.uk/our-work/low-carbon-future/decentralised-energy/index.aspx (accessed 16 January 2010).

Lovell, H. (2009) 'The role of individuals in policy change: The case of UK low-energy housing', *Environment and Planning C: Government and Policy* 27 (3): 491–511.

McFarlane, C. (2008) 'Governing the contaminated city: Infrastructure and sanitation in colonial and post-colonial Bombay', *International Journal of Urban and Regional Research* 32: 415–435.

McFarlane, C. and Rutherford, J. (2008) 'Political infrastructures: Governing and experiencing the fabric of the city', *International Journal of Urban and Regional Research* 32 (2): 363–374.

Monstadt, J. (2009) 'Conceptualizing the political ecology of urban infrastructures: Insights from technology and urban studies', *Environment and Planning A* 41 (8): 1924–1942.

Moss, T. (2003) 'Utilities, land-use change, and urban development: Brownfield sites as "cold-spots" of infrastructure networks in Berlin', *Environment and Planning A* 35 (3): 511–529.

OECD (2008) 'Competitive cities in a changing climate: An issues paper', in *Competitive Cities and Climate Change, OECD Conference Proceedings*, Milan, 9–10 October.

Raven, R. (2007) 'Niche accumulation and hybridisation strategies in transition processes towards a sustainable energy system: An assessment of differences and pitfalls', *Energy Policy* 35 (4): 2390–2400.

Raven, R. P. J. M., Heiskanen, E., Lovio, R., Hodson, M. and Brohmann, B. (2008) 'The contribution of local experiments and negotiation processes to field-level learning in emerging (niche) technologies: Meta-analysis of 27 new energy projects in Europe', *Bulletin of Science Technology Society* 28 (6): 464–477.

Romero Lankao, P. (2007) 'How do local governments in Mexico City manage global warming?', *Local Environment: The International Journal of Justice and Sustainability* 12 (5): 519–535.

Rutland, T. and Aylett, A. (2008) 'The work of policy: actor networks, governmentality, and local action on climate change in Portland, Oregon', *Environment and Planning D: Society and Space* 26 (4): 627–646.

Satterthwaite, D. (2008) 'Climate change and urbanization: Effects and implications for urban governance', United Nations Expert Group Meeting on Population Distribution, Urbanization, Internal Migration and Development, New York, 21–23 January, UN/POP/EGM-URB/2008/16.

Schot, J. (1998) 'The usefulness of evolutionary models for explaining innovation: The case of the Netherlands in the nineteenth century', *History and Technology* 14: 173–200.

Schreurs, M. A. (2008) 'From the bottom up: Local and subnational climate change politics', *Journal of Environment Development* 17 (4): 343–355.

Seyfang, G. (2009) 'Community action for sustainable housing: Building a low-carbon future', *Energy Policy*, online: http://dx.doi.org/10.1016/j.enpol.2009.10.027.

Seyfang, G. and Smith, A. (2007) 'Grassroots innovations for sustainable development: Towards a new research and policy agenda', *Environmental Politics* 16 (4): 584–603.

Smith, A. (2007) 'Translating sustainabilities between green niches and socio-technical regimes', *Technology Analysis & Strategic Management* 19: 427–450.

Smith, A., Voβ, J.-P. and Grin, J. (2010) 'Innovation studies and sustainability transitions: The allure of the multi-level perspective and its challenges', *Research Policy* 39 (4): 435–448.

Star, S. L. (1999) 'The ethnography of infrastructure', *American Behavioral Scientist* 43: 377–391.

Stern, N. (2006) *The Stern Review: The Economics of Climate Change*, London: HM Treasury.

Swyngedouw, E. (2006) 'Metabolic urbanization: The making of cyborg cities', in N. Heynen, M. Kaika and E. Swyngedouw (eds) *In the Nature of Cities: Urban Political Ecology and the Politics of Urban Metabolism*, London: Routledge.

Swyngedouw, E. and Heynen, N. C. (2003) 'Urban political ecology, justice and the politics of scale', *Antipode* 35 (5): 898–918.

UN (2008) *World Urbanization Prospects: The 2007 Revision*: New York: Department of Economic and Social Affairs, Population Division.

Unruh, G. C. (2002) 'Escaping carbon lock-in', *Energy Policy* 30 (4): 317–325.

While, A., Jonas, A. E. G. and Gibbs, D. (2009) 'From sustainable development to carbon control: Eco-state restructuring and the politics of urban and regional development', *Transactions of the Institute of British Geographers* 35 (1): 76–93.

Zérah, M.-H. (2008) 'Splintering urbanism in Mumbai: Contrasting trends in a multilayered society', *Geoforum* 39 (6): 1922–1932.

4 The carbon calculus and transitions in urban politics and political theory

Aidan While

Introduction

Concerns about energy security and limits on greenhouse gas emissions are leading to a new logic of what might be called 'carbon control' in spatial regulation (While *et al.* 2010). Carbon control in this sense signals the political imperative of reducing the use of fossil fuels as a first-order policy priority for governments and firms. Failure to do so carries the risk of penalties for non-compliance with carbon reduction targets, added costs arising from carbon taxes and rising energy prices, and falling behind in circuits of low carbon competitiveness. Low carbon restructuring is a concern at all levels of government, but the question addressed specifically in this chapter is what carbon control might mean for urban politics (and by extension, what urban politics might mean for low carbon transitions). For example, how might carbon management change the ways in which decision makers think about urban processes and urban policies? To what extent might a low carbon transition require new conceptions of the role of government, non-state actors and citizens, and their relationship with each other? And, not least, from the perspective of recent urban political theory, to what extent will seeing territory as a space of carbon flows reinforce or challenge dominant modes of (neo-liberal) urban management?

Politics matters in the low carbon transition. As has already been indicated in this book, there are multiple possibilities for ensuring low or zero carbon compliance, ranging from the technical (the retrofitting of urban infrastructure) to the social (facilitating or enforcing behaviour change), with a multitude of socio-technical possibilities in between. The mix of policy responses will vary between cities, but the processes and outcomes of low carbon restructuring will ultimately be determined by political choices about different pathways taken within the context of wider international and national carbon control regimes, and forged through compromise and negotiation between different interests at the urban scale.

This chapter is based around five premises. First, enabling a low carbon infrastructure will be an increasingly necessary task of urban governance, owing to more stringent carbon control measures and growing concerns about the future costs and security of supply of oil and gas. Second, with multiple possibilities for low carbon restructuring, urban leaders will be faced with a multitude of possible carbon management options cutting across all aspects of urban life. Third, by introducing a new calculus into decision making, the low carbon shift opens up possibilities for seeing cities differently, and especially for changing the ways in which things are valued in urban contexts. Fourth, cities will face differential challenges and opportunities in restructuring for a low carbon future, reflecting factors such as their position within inherited infrastructure networks, economic and political circumstances and the degree of political support for changes that might have a relatively long-term payback. Fifth, strategic decisions about low carbon restructuring are intimately bound up with questions of justice – that is, who gains and who loses – in the management of carbon control. Introducing the idea of the 'carbon calculus', the chapter begins by exploring the emerging regulatory logic of carbon management and its implications for spatial regulation.

Carbon control as a new calculus in state regulation

The international system for carbon control established in the 1997 Kyoto Protocol reflects a particular regulatory logic. First, there is a preference for flexible market-based solutions in which carbon units are priced (by the tonne of carbon) and then traded off within an agreed global emissions limit. The idea is that those who decrease their carbon emissions are rewarded by being able to sell their excess carbon credits, with continued carbon use from one source being offset by reductions elsewhere. The argument is that costing carbon in this way allows for trade-offs to be made about the most 'cost-effective' way of meeting targets within a clear limit (see Posner and Wiesbach 2010), but market-based solutions also allow governments to offset difficult political decisions about where cuts should be made. Second, international carbon control is based on a territorialised system of accounting in which the global limit for emissions is divided between nation-states. In effect, national governments are left to determine their own approach to meeting their target, including under the Kyoto Protocol the possibility that richer nations can gain credit for funding low carbon projects in developing nations. Third, the current logic of carbon regulation is based on 'downstream' costing, which targets the point at which carbon is emitted. This has implications for the ways in which carbon costs get located geographically, and thus who bears that cost. Fourth, there is a notional commitment to global equity, with advanced industrial nations taking on greater responsibility for cutting emissions. Carbon governance has thus resulted in a complex and multi-scalar approach for controlling emissions, which creates new remits for national governments while transferring regulatory power outwards to markets and non-nation state actors (Bailey 2007). The international carbon control regime established at Kyoto is by no means fixed, and its future direction will be determined by negotiation and struggle over future agreements and responses to the unknown and unintended consequences of the evolving carbon economy (see Boykoff *et al.* 2009).

In putting a value on something that is pervasive but as yet uncosted, carbon accounting necessarily introduces a new element into the decision-making calculus of firms, citizens and, not least, governments. Put simply, carbon emissions are now a commodity with a price that will need to be factored into all activities that have a fossil-fuel input, and this price will rise as increasingly stringent limits are passed on through market-based greenhouse gas reduction measures and carbon taxes (and the price of carbon will rise steeply if and when the costs of climate change become more apparent). The result is that those who are able to reduce their carbon dependence will be rewarded with lower costs, while high carbon users will pay more for their continued carbon fix.

Carbon management – that is, intervention to reduce carbon emissions entering the atmosphere – poses a series of challenges for governments in terms of encroaching on what have come to be regarded as individual personal freedoms, pushing through potentially unpopular infrastructure projects and/or eco-taxes, or finding the resources to invest in low carbon infrastructure. But carbon control is not simply about challenges for governments. With anxieties growing about the rising cost and security of gas and oil supplies (Newman *et al.* 2009), climate change arguments can be used to justify decarbonisation projects (including eco-taxes or neo-Keynesian green new deals) and legitimise the extension of state power and control (see Swyngedouw 2007). For some governments, low carbon restructuring will be seen as an opportunity to exploit competitive advantages in knowledge and technology.

Given the pervasiveness of carbon in all aspects of human activity, carbon control has profound and far-reaching implications. As Boykoff *et al.* (2009: 2299) argue, 'the carbon economy essentially and necessarily props up and connects the workings of our everyday lives to the national and global-level political economic architectures organizing contemporary human societies'. A carbon shift would imply a series of changes in investment and policy priorities, including vast investments in low carbon infrastructure to move away from dependence on fossil fuels; a raft of laws, regulations, taxes, subsidies and education programmes to incentivise and enforce new practices in building, mobility, production and consumption; and experimentation with new urban forms, food systems, and so on in order to ensure low carbon competitiveness. Much of this will be orchestrated by governments, though firms and citizens will also have direct financial incentives to make their own low carbon transitions.

More fundamentally, carbon management strategies open up possibilities for rethinking the value of things, representing not just an added financial calculation in cost–benefit analysis but also a 'reorganisation of economic principles' (Boykoff *et al.* 2009). In this context, carbon control can be seen as bringing a new set of 'calculative practices' (Miller 2004) into urban governance. As Mennicken *et al.* (2008: 5) point out, 'every mode of calculation produces a certain form of visibility which creates unique possibilities for intervention whilst displacing others', with new calculative practices helping to shape economic and social relations (see also Larner and Le Heron 2002; Henneberry and Roberts 2008). Miller (1994) suggests that there are three distinct dimensions to calculative practices: their technologies, their rationales and their relations with the wider economic domain. 'Technologies' in this sense refers to the methods and methodologies that are developed to facilitate decision making. The three dimensions of calculative practice are inextricably linked:

> The objectives of calculative technologies depend upon rationales, whilst the emergence of new rationales may prompt the development of new calculative methods. Similarly, changes in the conceptualisation of the economy may result from or result in changes in calculative technologies and their rationales. Existing understandings may become problematic when new calculative practices are introduced. The political economy of calculative practice may also be affected. Such practices are also vehicles for the exercise of power: partial and biased mechanisms that may further the interests of some classes or groups over others.
>
> (Henneberry and Roberts 2008: 1223–1224)

One important issue is that different sets of calculation will expose the established norms of a decision-making environment to 'alternative and competing norms' (Miller 1994: 13). Certainly, green political theorists have long argued that the environment tends to be undervalued in prevailing cost–benefit analysis, with decision making being based around a narrow economic rationality that privileges short-term paybacks over a consideration of longer-term social and ecological values (Scott Cato 2009). In this context, it would appear that carbon control represents a harder edge to state environmental regulation via non-negotiable target setting, and this would contrast with the fairly weak implementation of sustainability principles by Western governments. Carbon regulation certainly opens up fresh debate about policy choices and trade-offs at urban and regional scales.

However, established modes of calculation can be difficult to overturn and, as the policy history of sustainable development demonstrates, seemingly challenging ideas can get submerged or co-opted. Indeed, most writing to date has pointed to a less progressive side to the politics of carbon control in terms of reinforcing social and spatial inequalities and extending the reach of market environmentalism (Lohmann 2001, 2008). As Lohmann (2008: 3) argues, the emphasis on market solutions glosses over questions about how and where cuts are made, while Swyngedouw (2007) warns that the 'post-political' discourse of climate change can be used by governments to foreclose 'serious political questions about possible socio-environmental trajectories', for example by justifying socially or ecologically problematic low carbon transition fixes such as a resurgence of nuclear power or neo-Malthusian population controls. Others have been concerned that the broader social equity and ecological concerns of sustainable development might drop out of the equation as the instrumental goal of reducing carbon emissions comes to dominate (While *et al.* 2010).

Carbon control and the urban calculus

Since the late 1980s, the dominant strands of urban political theory have depicted a significant shift in the urban political calculus under the conditions of globalisation and neo-liberalisation. In the 'new urban politics' (Cox 1993), the focus of urban governance is seen as moving from a broad concern with the management of public goods to an overwhelming concern with economic competitiveness.

Opened up to international competition, cities have had to become more entrepreneurial in attracting (and retaining) mobile capital, diverting resources from social welfare to economic development (Harvey 1999; Jessop 1997; Peck and Tickell 2002). At the same time, urban service provision has become increasingly fragmented as infrastructure and services are separated out, packaged up and privatised, making it difficult for urban leaders to make connections over key aspects of the urban metabolism (Graham and Marvin 2001; Neuman and Smith 2010). There is, however, another strand of urban writing that points to a renewed concern in the post-industrial city with collective consumption, quality of life and proactive environmental policy as part of urban strategies to attract and retain high-value firms and workers (Jonas and While 2007; Jonas *et al.* 2010). There is also evidence of a renewed concern with 'ecological security' at the urban scale around energy, water and other resource inputs (Hodson and Marvin 2009). In these narratives, the agenda is not so much about driving down the costs of business as about investing in social and physical infrastructure in order to attract and retain higher-value firms, workers and residents. What links the various stories of neoliberal urban political transition is the issue of uneven spatial development. Not only have social and spatial divides increased within cities on the basis of income and opportunity, but divisions have also intensified between those cities able to pursue 'high-road' development strategies of smart growth and those on a 'low-road' trajectory of driving down the cost of business (see Malecki 2004). While this body of theory has largely reflected Western experience, it would appear to find broad application across the world.

If environmental policy has risen up the urban agenda in the post-industrial city, economic development has tended to dominate (While *et al.* 2004; Keil 2007). Aside from some notable exceptions, at the time of writing (early 2010), for the vast majority of cities the mitigation of climate change has not had a significant impact on hard decisions taken about investment or the allocation of resources. Although urban governments have enacted various carbon reduction initiatives on their own or through networks such as the Cities for Climate Change Protection or the US 'cool cities' movement (Bulkeley and Betsill 2003; Rutland and Aylett 2008; www.coolcities.us), carbon management has by and large been an activity at the margins of decision making. In short, carbon reduction strategies have not challenged the prevailing orthodoxy of urban policy. The reasons why carbon control has not started to bite at the urban scale are various, but in essence the 1997 Kyoto settlement did not translate into hard targets for reduction at international and national scale, with little or no penalty for non-compliance even for those countries that signed up to the agreement. Meanwhile, carbon economy mechanisms such as the EU Emissions Trading Scheme have so far not been traded at a level that would add significant costs along the commodity value chain (see Bailey 2007). As a result, much of the leading work by cities on climate mitigation has tended to be largely voluntaristic, pursued by activist authorities with a particular political commitment or rationale for intervention, including positioning the city for economic development opportunities (Rabe 2004, 2008). The following sections open up questions about what might happen as the combination of meaningful national territorial targets and market-based pressures associated with carbon accounting starts to impact more directly on the calculative practices of government.

The urban dimension I: the coming to ground of the carbon economy

As with other aspects of environmental policy, the urban politics of low carbon restructuring can be thought of as a negotiation between pressures and demands for carbon management and countervailing demands on urban leaders (While *et al.* 2004). Making an effective low carbon transition at the urban scale would include measures such as the organisation of energy-efficient land uses, investment in low carbon energy and transport infrastructures, more stringent regulations for energy conservation and renewable energy generation, engagement in renewable energy experiments, and work on changing the behaviour of citizens. But urban governments will also be under pressure to maintain existing levels of economic activity and there may be reluctance within the city to divert resources to carbon reduction actions. Nevertheless, the demands for effective urban carbon control regimes will increase as carbon accounting becomes part of the strategic context for decision making. This might happen in a number of ways, including:

- requirements on urban governments to comply with low carbon regulations set by higher levels of government in key areas of urban development and service delivery;
- pressures on urban governments from firms and residents to provide supporting infrastructure, policies and support for reducing their carbon footprints and minimising energy costs;
- the search for low carbon resilience in order to position the locality as a low carbon leader within future circuits of carbon competitiveness;
- competing for public and private investment in low-carbon demonstration projects;
- meeting sub-national territorial targets for emissions reduction, with financial penalties and other sanctions for non-compliance;
- increasing involvement of urban governments in carbon trading schemes.

The urbanisation of carbon control is therefore not simply about a simple cascade of policy from international levels (Bulkeley 2005), but rather it reflects the ways in which different sets of pressures come to ground and get mediated in different urban contexts. To some extent, the strategic context for urban action will be set by national modes of carbon management, not least in the division of responsibility for enforcing cuts between national and sub-national arms of the state. For example, sub-national carbon budgets or local authority cap and trade arrangements will only apply in countries where those approaches are promoted by higher levels of government as part of the mixed economy of carbon regulation. In the United Kingdom, for example, reducing carbon emissions is now one of the performance indicators used by central government to assess local authority performance, and from April 2010 all city governments have been enrolled in a mandatory local authority carbon trading scheme (see Sheffield City Council and Sheffield First Partnership 2009). Conversely, pressure from citizens might be greater in countries where personal carbon trading is introduced or where energy prices are allowed to rise without large levels of state subsidy. In short, different national carbon control and energy management regimes enable and constrain different possible responses in different urban contexts, though some cities might be able to transcend their national political context by enrolling in wider international networks, including cap and trade schemes (Hodson and Marvin 2009). What is certain is that no city will be immune to the pressures and demands exerted by the carbon economy.

The urban dimension II: seeing the city as a space of carbon flows

A key question for the politics of low carbon transition is how carbon accounting might alter the calculus of urban politics. At one level, this is about factoring the financial costs of future low carbon regulation into urban strategies. Thus, a study on the possible future impact of EU and UK carbon legislation on the economy of the city of Manchester has outlined a stark choice of either innovating and adapting existing industries or lowering output in order to reduce emissions. Moreover, the prognosis is that whatever changes are introduced, there will need to be a profound shift in the city's economic base, as some of the current economic strengths will become unviable in a low-carbon future (Deloitte 2008). As the Manchester study makes clear, strong leadership is required to ensure that the city is able to 'enhance its competitive advantage over those cities that are slower to adapt' (ibid.: 7).

One consequence of the carbon calculus is that urban leaders will have to take a more proactive role in managing flows of carbon into, within and out of the city – in effect, seeing the city as a space of carbon flows. Moreover, if urban planners have been described as 'doctors of space' (Lefebvre 1991: 99), urban managers will increasingly be seen as 'doctors of carbon flows' charged with securing a carbon fix within limits set by targets and the cost of carbon. In this context, Rutland and Aylett (2008) have drawn attention to the role of the carbon inventory as a technology for rendering explicit carbon outputs in a given territory, with these inventories becoming ever more detailed and forming the basis for modelling low-carbon options in the future. For example, the UK city of Sheffield's urban leadership has set a target to reduce the city's carbon emissions by 30 per cent as compared with 2005 figures by 2020, with a 60 per cent reduction by 2050 (Sheffield City Council and Sheffield First Partnership 2009). Like leaders in many cities, Sheffield's civic leaders have undertaken a carbon inventory and commissioned modelling studies on possible approaches to carbon cost saving. What is interesting is the range of 'social' options that are included within the low carbon calculus, including 'investment in a city-wide affordable warmth programme', 'a shift to vegetarian and organic diets', 'reductions in food miles', 'improved car engine efficiency', 'a modal shift in transport' and 'different options for renewable energy take-up', among others (Roberts *et al.* 2009). In effect, the Sheffield study connects different spheres of low carbon action and places them within a single decision-making space, allowing for comparison of the cost-effectiveness of different present and future options in order to meet a given reduction goal. In other words, seeing the city as a space of carbon flows begins to present a range of different sets of options for managing the city as an integrated socio-economic and socio-technical entity.

By defining and redefining the object of regulation, emissions inventories shape the sorts of connections or disconnections that are likely to be made around carbon governance. In Portland, Oregon, for example, the city's emissions inventory was limited to actions that 'could be influenced by government' – that is, that could be governed – which excluded emissions from air travel, long-distance transportation of imported commodities and the outsourced emissions of locally consumed commodities (Rutland and Aylett 2008: 640). The implications of seeing the city as a space of carbon flows are extended significantly if, as the Sheffield study anticipates, carbon control moves from downstream regulation (i.e. mainly about outputs) to making citizens and cities accountable for the embodied carbon inputs of the things they consume (Davis and Caldeira 2010).

The urban dimension III: political transitions

If carbon management is linked to fundamental shifts in the basis of production and consumption, it is likely to have significant impacts on the ways in which cities are governed and regulated. For example, by altering the basis of strategic calculation, carbon accounting would appear to open up 'alternative' possibilities for economic development in cities, especially given the narrow calculus of the competition state. In some cities, this might simply be about political support for hitherto marginalised environmental policy measures such as constraints on car use, decentralised energy systems, subsidies for pro-environmental behaviour, and so on. More radically, seeing the city differently, as a space of carbon inputs and outputs, might encourage city leaders to explore possibilities for protecting and developing urban carbon sinks. New relationships might need to be formed at the regional and city-regional scale, or between networks of cities, based on carbon interdependencies. Going further, commentators such as Peter North (2010) have seen opportunities for more transformative green restructuring around low-growth, alternative-growth or localisation strategies, such as those pursued by the United Kingdom's Transition Towns movement, with its emphasis on local empowerment and taking control of local food and power supply (Hopkins 2008). Whether these alternative opportunities are open to all places, or all communities, is less clear.

A key strand of urban political theory over the past twenty years has been a concern with the differential capacity of localities to influence their economic fortunes, reflecting a shift in emphasis from the exercise of power over citizens to the ability of growth regimes to orchestrate post-industrial economic transitions (see, for example, Harding 1996; Wood and Valler 2004). The era of carbon control raises similar questions about the capacity of cities to make low carbon transitions, albeit on the basis of a slightly different set of criteria, including levels of carbon dependence (in the economy, infrastructure, land-use patterns, etc.), the costs of low carbon retrofitting and restructuring, the degree of political and public support for low carbon measures, and, not least, the competing costs of climate change adaptation. Any low carbon transition will also have to take into account the various other pressures and demands on urban leaders, including the ways in which they have to compete for investment. In short, low carbon transitions present variable challenges and opportunities for localities, and cities and regions have differential capacities and resources to make a low carbon transition.

While carbon control might change the calculus of investment and management priorities within cities, it is perhaps unlikely to challenge the logic of inter-urban competition, and may intensify divisions between cities. On the basis of existing trends, wealthier cities will find it easier to engage in experiments with low carbon energy systems, invest in low carbon transportation systems, gain public support for low carbon measures or offset their carbon use by supporting low carbon projects elsewhere. Predominantly service-sector economies may prove more amenable to a low carbon shift, while economically marginal areas would continue to be locked into the fossil fuel dependence of the residual carbon economy (Innovas 2009). There is therefore a question of distributional politics – or carbon justice – at the heart of urban low-carbon transitions, and this includes questions about the division of responsibility for carbon governance at city-region or regional scales, given concerns about carbon free-riding by suburban residents (Glaeser 2009). While the carbon regulation could be used to equalise spatial inequalities and promote alternatives, another scenario is that the carbon calculus will lead to a new urban protectionism as leading city-states seek to protect their carbon fix through, *inter alia*, experiments in the decentralisation of energy-supply systems or greater selectivity about the sorts of activities and people that are allowed to use valuable energy resources.

As the politics of carbon control intersects more closely with urban politics, questions of distributional politics also become important *within* cities. On the face of it, establishing a price for carbon emissions offers socially progressive possibilities for taxing wealthier high carbon users and rewarding poorer social groups, who in theory emit less carbon but suffer most from the health effects and economic inequalities resulting from carbon use (Seyfang 2007). However, carbon output differentials between the poor and the wealthy may quickly reduce as richer residents invest in low carbon solutions and benefit from the payback such investment offers. Moreover, in their role as carbon police, urban governments will reserve the right to reward and incentivise certain behaviours (and certain groups) while punishing or marginalising others. Seeing the city as a space of carbon flows would allow urban leaders to offset some of the carbon costs of the economic growth machine by disciplining and punishing carbon users in other spheres.

Conclusions

The aim of this chapter has been to explore some of the possible implications of low carbon transitions for urban politics. In particular, the focus on calculative practices has sought to provide a conceptual frame in which possible low carbon transitions can be located within the political calculus of urban governance. The central premise is that for all cities, controlling carbon flows will become an increasingly important part of the 'calculus' of urban politics, opening up new ways of seeing and valuing different elements of the urban assemblage. Although carbon control will present different mixes of challenges and opportunities for different cities, no city will be immune from the workings of the carbon economy.

If there are strong political-economic incentives for governments to move forward on carbon control, there is much to be played out in terms of low carbon restructuring, and this will include factors such as technological innovation, the future price of oil and wider economic circumstances. In short, there are multiple possible low carbon futures, all of which have different implications for thinking about urban management. What this chapter has sought to highlight are some of the difficult political choices that will underpin decisions about low carbon urban transitions, including the availability of funds for investment in low carbon collective provision, competing pressures and demands for what resources are available as urban leaders seek to balance short- and longer-term objectives, and potential public and political resistance to change. It might also be added that urban carbon management entails a complex set of issues around the timing of low carbon projects, given the likelihood of technological innovation and changes in the ways in which carbon emissions are regulated. Meanwhile, for most cities adaptation to climate change itself will require significant level of investment, and in some cases this will mean that the city can no longer function as a meaningful economic entity. Within this context, seeing the city as a space of carbon flows certainly opens up opportunities for experimentation with what might be regarded as being 'alternative' urban strategies, and perhaps alternative urbanisms, for urban leaders and their citizens, including 'counter-hegemonic' options that challenge the social and environmental injustices associated with neo-liberalism. But carbon control also brings with it the possibility of new forms of carbon inequality as urban low carbon transitions are determined by the economic bottom line of carbon competitiveness.

References

Bailey, I. (2007) 'Climate policy implementation: Geographical perspectives', *Area* 39 (4): 415–417.

Boykoff, M., Bumpus, A., Liverman, D. and Randalls, S. (2009) 'Theorizing the carbon economy: Introduction to the special issue', *Environment and Planning A* 41 (10): 2299–2304.

Bulkeley, H. (2005) 'Reconfiguring environmental governance: Towards a politics of scales and networks', *Political Geography* 24 (8): 875–902.

Bulkeley, H. and Betsill, M. (2003) *Cities and Climate Change: Urban Sustainability and Global Environmental Governance*, London: Routledge.

Cox, K. (1993) 'The local and the global in the new urban politics: A critical view', *Environment and Planning D: Society and Space* 11: 433–448.

Davis, S. J. and Caldeira, K. (2010) 'Consumption-based accounting of CO_2 emissions', *Proceedings of the National Academy of Social Sciences of the United States of America* 107 (19): 5687–5692.

Deloitte (2008) *'Mini-Stern' for Manchester: Assessing the Economic Impact of EU and UK Climate Legislation on the Manchester City Region and the North West*, London: Deloitte. Online, available at: www.deloitte.com/assets/Dcom-UnitedKingdom/Local%20Assets/Documents/UK_GPS_MiniStern.pdf (accessed 26 March 2009).

Glaeser, E. L. (2009) 'Green cities, brown suburbs', *City* 19 (1). Online, available at: www.city-journal.org/2009/19_1_green-cities.html (accessed 2 February 2010).

Graham, S. and Marvin, S. (2001) *Splintering Urbanism: Networked Infrastructures, Technological Mobilities and the Urban Condition*, London: Routledge.

Henneberry, J. and Roberts, C. (2008) 'Calculated inequality? Portfolio benchmarking and regional office property investment in the UK', *Urban Studies* 45: 1217–1241.

Hodson, M. and Marvin, S. (2009) 'Urban ecological security: A new urban paradigm?', *International Journal of Urban and Regional Research*, 33 (1): 193–215.

Hopkins, R. (2008) *The Transition Handbook: From Oil Dependency to Local Resilience*, Totnes, UK: Green Books.

Harding, A. (1996) 'Is there a new "community power" and why should we need one?', *International Journal of Urban and Regional Research* 20 (4): 637–655.

Harvey, D. (1996) *Justice, Nature and the Geography of Difference*, Oxford: Blackwell.

Innovas (2009) *Low Carbon and Environmental Goods and Services: An Industry Analysis*, report to the Department for Business Enterprise and Regulatory Reform, Winsford, UK: Innovas.

Jessop, B. (1997) 'A neo-Gramscian approach to the regulation of urban regimes: Accumulation strategies, hegemonic projects and governance', in M. Lauria (ed.) *Reconstructing Urban Regime Theory: Regulating Urban Politics in a Global Economy*, Thousand Oaks, CA: Sage.

Jonas, A. E. G. and While, A. (2007) 'Greening the entrepreneurial city: Looking for spaces of sustainability politics in the competitive city', in R. Krueger and D. Gibbs (eds) *The Sustainable Development Paradox: Urban Political Economy in the United States and Europe*, New York: Guilford Press.

Jonas, A. E. G., While, A. and Gibbs, D. (2010) 'Managing infrastructural and service demands in new economic spaces: The new territorial politics of collective provision', *Regional Studies* 44 (2): 183–200.

Keil, R. (2007) 'Sustaining modernity, modernizing nature: The environmental crisis and the survival of capitalism', in R. Krueger and D. Gibbs (eds) *The Sustainable Development Paradox: Urban Political Economy in the United States and Europe*, New York: Guilford Press.

Larner, W. and Le Heron, R. (2002) 'The spaces and subjects of a globalising economy: A situated exploration of method', *Environment and Planning D: Society and Space* 20 (6): 753–774.

Lefebvre, H. (1991) *The Production of Space*, Oxford: Blackwell.

Lohmann, L. (2001) *Democracy or Carbocracy? Climate Trading and the Future of the Climate Debate*, Sturminster Newton, UK: The Corner House.

Lohmann, L. (2008) 'Carbon trading, climate justice and the production of ignorance: Ten examples', *Development* 27: 1–7.

Malecki, E. (2004) 'Jockeying for position: What it means and why it matters to regional development policy when places compete', *Regional Studies* 38: 1101–1120.

Mennicken, A., Miller, P. and Samiolo, R. (2008) 'Accounting for economic sociology', *Economic Sociology: The European Electronic Newsletter* 10 (1): 3–7.

Miller, P. (1994) 'Accounting as social and institutional practice', in A. G. Hopwood and P. Miller (eds) *Accounting as Social and Institutional Practice*, Cambridge: Cambridge University Press.

Miller, P. (2004) 'Governing by numbers: Why calculative practices matter', in A. Amin and N. J. Thrift (eds) *The Blackwell Cultural Economy Reader*, Oxford: Blackwell.

Neuman, M. and Smith, S. (2010) 'City planning and infrastructure: Once and future partners', *Journal of Planning History* 9 (1): 21–42.

Newman, P., Beatley, T. and Boyer, H. (2009) *Resilient Cities: Responding to Peak Oil and Climate Change*, Washington, DC: Island Press.

North, P. J. (2010) 'Eco-localisation as a progressive response to peak oil and climate change: A sympathetic critique', *Geoforum* 41 (4): 585–594.

Peck, J. and Tickell, A. (2002) 'Neoliberalizing space', *Antipode* 34 (3): 380–404.

Posner, E. A. and Wiesbach, D. (2010) *Climate Change Justice*, Princeton, NJ: Princeton University Press.

Rabe, B. G. (2004) *Statehouse and Greenhouse: The Emerging Politics of American Climate Change Policy*, Washington, DC: Brookings Institution Press.

Rabe, B. G. (2008) 'States on steroids: The intergovernmental odyssey of American climate policy', *Review of Policy Research* 25 (2): 105–128.

Roberts, G., Hayes, D., Hunt, H. and Patient, J. (2009) 'A green new deal for Sheffield: From global challenges to local solutions', Conference Report. Online, available at: www.sheffield.gov.uk/your-city-council/council-meetings/cabinet/agendas-2009/agenda-12th-august-2009 (accessed 3 February 2010).

Rutland, T. and Aylett, A. (2008) 'The work of policy: Actor networks, governmentality, and local action on climate change in Portland, Oregon', *Environment and Planning D: Society and Space* 26: 627–646.

Scott Cato, M. (2009) *Green Economics*, London: Earthscan.

Seyfang, G. (2007) 'Personal carbon trading: Lessons from complementary currencies'. Online, available at: www.uea.ac.uk/env/cserge/pub/wp/ecm/ecm_2007_01.pdf (accessed 10 January 2010).

Sheffield City Council and Sheffield First Partnership (2009) 'Reducing Sheffield's carbon emissions: A strategic framework', Sheffield: Sheffield City Council and Sheffield First Partnership. Online, available at: www.sheffieldfirst.net/EasySite/lib/serveDocument.asp?doc=178540&pgid=139695 (accessed 10 December 2009).

Swyngedouw, E. (2007) 'Impossible "sustainability" and the postpolitical condition', in R. Krueger and D. Gibbs (eds) *The Sustainable Development Paradox: Urban Political Economy in the United States and Europe*, New York: Guilford Press.

While, A., Jonas, A. E. G. and Gibbs, D. C. (2004) 'The environment and the entrepreneurial city: Searching for a sustainability fix in Leeds and Manchester', *International Journal of Urban and Regional Research* 28: 549–569.

While, A., Jonas, A. E. G. and Gibbs, D. (2010) 'From sustainable development to carbon control: Eco-state restructuring and the politics of urban and regional development', *Transactions of the Institute of British Geographers* 35 (1): 76–93.

Wood, A. and Valler, D. (eds) (2004) *Governing Local and Regional Economies: Institutions, Politics and Economic Development*, Aldershot, UK: Ashgate.

5 Can cities shape socio-technical transitions and how would we know if they were?

Mike Hodson and Simon Marvin

Introduction

Cities have been transformed over the past two decades in ways that have been characterised as their 're-emergence', 'renaissance' and 'entrepreneurialism' (see Brenner 2004; Harvey 1989; Jessop 1997). Essential infrastructures of energy, water, waste and transport have been fundamental in supporting this 're-emergence'. Yet until recently, the provision and organisation of these critical infrastructures were largely perceived unproblematically, and taken for granted as primarily engineering challenges and administrative issues (Graham and Marvin 2001). More recently, however, a series of economic, ecological, population and institutional constraints have produced new challenges and pressures on urban growth and the management of cities' essential infrastructures. Furthermore, the push for 'competitiveness' and place-based competition is occurring while established energy, water, waste and food resources that underpin economic growth are increasingly constrained and the basis of recontinued geopolitical struggle and subject to securitisation (Dalby 2007). Questions about the security of ecological resources and the implications of climate change have become increasingly internalised and intertwined with national states' priorities and responsibilities for social welfare and economic competitiveness. These are also increasingly becoming issues at an urban scale (Hodson and Marvin 2009a).

Although these new pressures are generic, their presence is experienced differentially across cities and also within cities. In particular, we are interested in the urban responses to these pressures and what these mean for future configurations of the social and technical organisation of infrastructure systems in cities (Hodson and Marvin 2010). Attempts to develop systematic change meet the historical legacy of infrastructural systems that were frequently developed a century and more ago in many Western contexts and whose capacity to be managed in an integrated way is now highly fragmented. Critically, the (variable) privatisation and the liberalisation of many infrastructures and the opening up to competition of infrastructure provision mean that a wide range of distributed stakeholders and social interests are now involved in the functioning of socio-technical infrastructure systems (Graham and Marvin 2001). The functioning of infrastructures is often seen from very many different viewpoints and positions –

including utilities, local authorities, regulators, consumers, citizens, businesses and so on – in respect of different issues – such as economic growth, climate change, resource consumption – at many different levels – supranational political institutions, national government, regional agencies, local authorities, businesses, households, and so on.

'Effective' urban responses to these pressures are thus predicated on multiple challenges, multiple actors and multiple levels that require effective coordination to inform the effective reconfiguration of infrastructure systems. Urban infrastructure transitions require new forms of knowledge and capacity to be produced, communicated and deployed. Critical to the different ways in which cities constitute responses is the development of new intermediaries that sit between production and consumption interests seeking to (selectively) develop the knowledge, expertise and capability to functionally manage systemic transitions in urban infrastructure. In this chapter, therefore, we critically examine the underexplored relationship between the multilevel perspective (MLP) on socio-technical transitions (Geels 2004; Elzen *et al.* 2004) and cities, particularly world cities (Sassen 2001), where pressures to undertake socio-technical transitions in infrastructure systems, we argue, are particularly pronounced and there are expectations, aspirations and plans to undertake purposive socio-technical transitions. In doing this, we specifically examine the complementarities and the shortcomings of the MLP's conceptualisation of long-term transitions in socio-technical systems with a relational and multilevel-governance-informed understanding of the production and re-production of cities. Addressing these issues requires that we understand the generic pressures contemporary cities are faced with, outline the dominant urban strategic responses to these pressures, and ask how they are constituted and with what consequences. It also means that we do not merely accept the 'obviousness' of these strategies. We do this in order to more clearly conceptualise the role of cities within systemic socio-technical transitions and to enhance our understanding of the social processes of shaping urban infrastructure networks.

Emerging purposive and systemic transitions in cities' infrastructures

Increasingly, it is particular coalitions of social interests (usually involving urban political, policy and business elites and decision makers, and, to a lesser extent, including NGOs and environmental justice groups), often working within and/or through the largest and most powerful cities (e.g. London, New York, cities of the C40), that are developing a more strategic orientation towards critical questions about future resource requirements and infrastructural implications. The strategic response is leading to the development of new styles of infrastructure development that privilege particular spatial and socio-technical configurations of infrastructure. In a period of resource constraint and climate change, cities are beginning to translate their strategic concern about their ability to guarantee resources into strategies designed to reshape the city and its relations with resources and other spaces in three critical ways.

First, the strategic protection of world cities. Central to the strategies of 'cities' is investment in generating a systemic understanding of the city-specific and long-term effects of climate change, especially in relation to flood risk and temperature rise and the development of systemic responses through strategic flood protection, green infrastructure and retrofitting to deal with increased temperatures. Given the significant uncertainties about precisely how climate change-induced effects will impact on particular places, a premium is increasingly placed on how 'cities' develop the networks and relationships to develop new knowledge and intelligence (Luke 2003). Significantly, it is in the context of world cities that the most significant progress is being made in generating such contextual knowledge, where climate change may shape the context within which the city develops but also, more critically, how longer-term access to critical resources and materials may be reshaped. Particularly emblematic of such approaches is the Thames Estuary 2100 study, which aims to determine the appropriate level of flood protection needed for London and the Thames Estuary for the next hundred years.

Second, the construction of more self-reliant urbanism. Cities have usually sought to guarantee their reproduction by seeking out resources and sinks from locations usually ever more distant. Yet this traditional approach is now being challenged as cities seek to re-internalise resource endowments and create the recirculation of wastes as they withdraw from reliance on international, national and regional infrastructures. Key examples are New York's strategy of energy independence, the recent doubling of decentralised energy targets in London, and Melbourne's development of renewable-powered desalination. Alongside such strategies, cities are attempting to reduce reliance through water and energy conservation and waste minimisation schemes, and by developing pricing mechanisms for car-based mobility and reducing reliance on 'external resources'. The new socio-technical configuration is similar to the concept of the autarkic or autonomous city. This new strategy is no longer based on stretching networks to search out resources more distant from cities but on a strategy of withdrawal and seeking greater independence through developing local resources. Such a socio-technical strategy combines both ecological and security priorities in a new strategy of attempts to guarantee secure urbanism and resilient infrastructure.

Third, the development of new networks of global urban agglomerations. New networks of social interests speaking on behalf of selected cities are constituting their interests in initiatives like the C40 and the Clinton Climate Initiative. Collectively, these cities are working together to develop self-reliant urbanism. For example, they are developing common measurement tools so that cities can establish a baseline on their greenhouse gas emissions, track reductions and share best practice to inform mitigation and evaluation activities. By mobilising and networking expert assistance, the self-stated objective is to help cities develop and implement programmes that will lead to reduced energy use and lower greenhouse gas emissions in areas including building efficiency, cleaner transport, renewable energy production, waste management, and water and sanitation systems. These cities are collectively building new global urban agglomerations of new mobility systems. At the same time as focusing on the protected space, bounding and enclosing resources, they seek to guarantee intra-city and inter-world city mobility through new technologies including pricing, transport informatics and new fuel systems based on hydrogen, biofuels or complex hybrids.

Socio-technical transitions, the MLP and cities

'World cities' are seeking to respond strategically to generic pressures by developing managed systemic change in the socio-technical organisation of key aspects of their infrastructure. Now we will examine how the multilevel perspective on socio-technical transitions can help us to understand the role of cities, identify the critical gaps that are not addressed and assess where the MLP would benefit from additional development. In doing this, we specifically examine the complementarities and the shortcomings of the MLP's conceptualisation of long-term transitions in socio-technical systems with a relational and multilevel-governance-informed understanding of the production and re-production of cities.

MLP and system innovation

The MLP provides an ambitious attempt to develop understanding of 'system innovation' (Geels 2002a, b). In doing this, it situates technological transformation in relation to wider socio-political economic 'systems'. Analytic understanding of these processes of 'system innovation' and socio-technical transitions is predicated on an interrelated three-level framework of landscape (macro), regime (meso) and niche (micro). The concept of 'landscape' is important in the MLP in seeking to understand the broader 'conditions', 'environment' and 'pressures' for transitions. The landscape operates at the macro level, focuses on issues such as political cultures, economic growth, macroeconomic trends, land use, utility infrastructures, and so on (Geels 2002b: 369), and applies pressures on existing socio-technical regimes, creating windows of opportunities for responses (Geels and Schot 2007). Socio-technical regimes situate existing or incumbent technologies within a 'dynamically stable' (ibid.) configuration of institutions, practices, regulations, and so on, where configurations impose a logic, regularity and varying degrees of path dependencies on technological change. The emphasis on regimes, therefore, highlights the enablement and constraints on new technologies breaking through where 'reconfiguration processes do not occur easily, because the elements in a socio-technical configuration are linked and aligned to each other' (Geels 2002a: 1258). The idea of socio-technical niches, which operate at a micro level, is one of 'protected' spaces, usually encompassing small networks of actors learning about new and novel technologies and their uses, and seeking to get new technologies onto 'the agenda', that attempt to keep alive novel technological developments which otherwise might be 'unsustainable' (Geels 2002b; Hoogma *et al.* 2002).

Adrian Smith and colleagues (Smith *et al.* 2005; Berkhout *et al.* 2003), while acknowledging the strengths of transitions approaches, make a thoughtful and constructive contribution to this debate. They question the view that regime change begins in niches and works upwards, arguing that this underplays the importance of the relationship between landscape pressures and regimes. In particular, they characterise regime change as being predicated on the ways in which shifting pressures impinge on a regime and the extent of the coordination of responses to these pressures from both inside and outside the regime. In doing this, they open up the issue of the governance (rather than government) of regime transformation in respect of agency and intervention in relation to both landscape and regime. They point out that landscape pressures can be articulated differently both in very general terms and in relation to specific regimes. It is not only the articulation of these pressures but also the relationships, resources and their levels of coordination that constitute a response to these pressures. Consequently, efforts to establish and enact capacity can be seen as the governance of regime transformation, which addresses the extent to which regime transformation is purposively informed or the outcome of historical processes – in effect, the balance of the relationship between structure and agency. Highlighting the context of the regime in transitions, the importance of governance processes and the coordination of capacity opens up the possibilities for understanding a variety of transition pathways and in doing so raises the issue of the extent to which pressures on the regime are responded to through resources and relationships incumbent within the regime or co-opted from outside the regime.

The absence of cities in multilevel transitions approaches

The multilevel perspective on transitions thus highlights the importance of the nested interrelationships of wider landscape 'environments', the stability and interrelationships of regimes and the innovative possibilities of niches. It outlines a way of thinking about the relationships, resources and practices, including technologies, institutions, skills and so on, that sustain existing configurations and regimes, but also addresses processes of adapting and evolving such a regime in relation to 'pressures' for, and contexts of, new technological possibilities and innovations. Yet despite an impressive breadth of focus on substantive areas as varied as transport, energy, water, waste and food systems (Hoogma *et al.* 2002; Verbong and Geels 2007; van der Brugge and Rotmans 2007; Geels 2005; Green and Foster 2005), frequently within a context of wider transitions to sustainability (Elzen *et al.* 2004) and often with a focus on institutional and governance issues in relation to transitions (Voβ *et al.* 2006), spatial scale frequently remains implicit or underdeveloped in the MLP and transitions approaches generally. The consequence of this is that there is an absence of clarity about 'where' transitions take place.

Transitions approaches have been limited in focusing on spatial scales aside from the national level. In particular, they have said little about cities and what the multilevel perspective on systemic transitions can contribute to understanding urban social-technical transitions. Usually there is at least an implicit emphasis on 'national'-scale transitions that requires understanding of particular socio-technical national contexts and their historical, institutional and policy contexts, and also the mechanisms, politics and processes through which attempts are made to steer transitions. But within the national view of transitions, the role for sub-national scales – that is, regions, cities, localities and so on – is largely absent. This is surprising, given that more than half the world's population now lives in cities (United Nations 2006), where cities are viewed as sites of intense economic activity and where recent IEA figures suggest that most carbon-related emissions from energy use come from cities (IEA 2008). Yet if the focus is on national transitions, we can then envision a number of different roles for sub-national scales and how they potentially interrelate with wider societal transitions. An urban transition can form a variety of different types of relationships with national transitions. Central to this potential is the relative positioning of cities in terms of their position in urban hierarchies and governance capacity, which means that cities have differential capacity to be either 'shapers of' or 'shaped by' national transitions:

- Initially, then, do we conceive of the city as a scale for 'receiving' national transitions that are then 'implemented' in local context?
- If this is the case, can different configurations of social interests at the urban scale mediate national transitions – that is, 'accelerate', 'reshape' or even 'disrupt' the implementation of national transitions in their local context?
- If urban social interests can mediate national transitions, can they then develop further capacity and capability to envision and enact their own locally developed urban transitions that are relatively distinct from national transitions?
- Is it then even possible to conceive of cities developing transition initiatives at urban level that are 'taken up' by the national context and reincorporated into new national transitions and then cascaded downwards onto cities (see Hodson and Marvin 2009b)?

Yet the role of cities in transitions approaches is consequently uncertain, fragmented and often implicit (see Monaghan *et al.* 2009). But this then still raises the issue of where cities 'fit' within the multilevel perspective and, in particular, where cities sit within the landscape–regime–niche hierarchy. Indeed, can they be encompassed by both regime and niche?

Understanding the role of cities in a multilevel transitions perspective needs also to take seriously multilevel governance (see Bache and Flinders 2004) and different scales of action. Agency at the level of the city cannot be reduced to understanding the variety and coalitions of actors (e.g. local authorities, mayors, universities, local economic actors, etc.) attributed to work at this scale. It also involves, and requires understanding of, the influence of actors at national and supranational scales of action who influence, both intentionally and through unintended consequences, action at a city scale through the production of new state spaces (Brenner 2004). To put it another way, there are multiple scales of governance action with differing sets of power relations operating in the relationships between these scales of action, and these power relations between different scales of action are variably constituted and organised in respect of different cities. Questioning critically these relationships between scales allows us to conceive of cities not merely as sites for receiving transition initiatives but also potentially as contexts for more purposive urban transition.

Understanding purposive urban transitions: a framework

Critically for us, then, the question is, how do we address the relative neglect of cities in the MLP? Here we seek to address this critical gap by developing a framework for understanding the distinctiveness of purposive urban transitions.

A series of contemporary landscape pressures – including resource scarcity, responses to climate change, maintaining economic competitiveness, the struggle between public and private – while being exerted generically, are being articulated more specifically in relation to cities, and in particular in relation to world cities (see Sassen 2001; and also the collection by Brenner and Keil 2006). This, of course, raises questions as to who is doing this and who is claiming to speak on behalf of these cities. As our series of propositions outlined above show, these pressures are becoming manifest in strategies of infrastructural transformation. What is particularly interesting here is that territorial priorities at the scale of the city – such as economic growth targets, aspirations to reduce carbon emissions – are becoming strategically intertwined with the reconfiguration of socio-technical infrastructure systems that may, or more often may not, be organised at the scale of the city. That is to say that urban governance networks (municipal and local policy makers and officials in particular) may sit outside of socio-technical regimes but to achieve territorial priorities need to gain degrees of influence and control over regimes.

The separation is not always a neat one: some members of urban governance networks, for example, also have roles within existing socio-technical regimes. The distinction we are making here is that frequently urban governance networks are organised as urban growth regimes. It is from this position of seeking to maintain and promote urban growth in an era of effectively responding to climate change and resource constraint that urban governance networks increasingly seek to become involved in attempts to reshape and control infrastructure networks at an urban scale. How this happens is another matter. This may be through urban governance networks seeking to reconfigure a regime through fundamentally influencing the shape of an existing regime from within; or, alternatively, through the cultivation of an urban niche or series of niches that potentially constitute a nascent urban regime; or where these urban niches contribute to the transformation of an existing socio-technical regime through their upscaling from an urban context.

Consequently, territorial governance priorities increasingly require degrees of control and influence over energy, water, waste and transport regimes. In terms of a purposive transition, the issue this raises is the extent to which urban territorial governance priorities can effectively be coordinated and aligned with the priorities and social interests that constitute socio-technical regimes. With the notion of transition pathways in mind, the mutual constitution of urban territorial governance priorities and socio-technical regimes may vary across different cities. This is particularly so given (1) the different histories of the socio-technical organisation of regimes and their relationships to cities, and (2) the regulatory states and multilevel governance relationships, both 'internal' and 'external' to cities, which relationally constitute a view of, and a claim on, 'the city'. In respect of the former, there may or may not be or have been historically a strong role for the municipal state and metropolitan agencies in the organisation and provision of energy, water, waste and transport systems. There is the likelihood for a variable role both between cities and at different points in time. Relating to the latter, cities are enmeshed more or less strongly in multilevel governance relationships where, for example, cultures of centralisation (as in the United Kingdom) or federalism (as in Germany) condition the nature of multilevel governance relationships.

The issue here is the degree to which there is 'separation' or alignment between urban governance network priorities and socio-technical regime priorities. To use the language of the MLP, it is the extent to which the territorial priorities of an urban governance network – and the social interests that produce them – are able to actively manage socio-technical regime change. Linked to this is the need for an understanding of the existing organisation of socio-technical regimes and the degree of involvement of urban governance representatives within them; but also the necessity for understanding the potential for mutually reconstituting urban governance networks and socio-technical regimes. Yet just as socio-technical regimes are constituted by complicated and often obdurate relationships, resources, regulations, artefacts, and so on, so too are urban governance networks.

In short, urban responses to these pressures will be variable. Cities and geographies will not only experience these challenges differently but also have historically organised infrastructure provision that may differ, and a variable capacity at an urban scale to respond to the emerging pressures. Three issues are important to understand whether an envisaged transition is predicated on a distinctly urban socio-technical regime, the degree of regime change required, the urban governance capability to enact such changes, and the ways in which there would be common understanding that it had been done. These are: (1) how pressures are experienced and perceived in a particular city and by whom, and how this translates into a shared understanding of an urban socio-technical transition; (2) the current and historic organisation of infrastructure in relation to a city and the level of capacity and capability to develop and operationalise this shared understanding processually; and (3) the degree of learning that takes place within and about the urban transition.

Shared visions of urban transitions?

In thinking through what an urban transition would look like, it is necessary to understand the extent to which there is a common and shared understanding among a wide range of social interests that produce territorial priorities and those of socio-technical regimes of energy, water, waste and transport infrastructures. 'Visions', which are a central part of prospective transitions management approaches (Kemp and Loorbach 2005; Rotmans *et al.* 2001), offer the potential to both constitute and to present a shared understanding of territorial and regime interests. They provide an ongoing locus for the development of shared understanding but also require as a precondition the coalescence of urban governance networks and socio-technical regime interests to do so. In this respect, visions are normative, but they are also dynamic and the subject of negotiation, struggle and exclusions. Visions of an urban transition may bring an understanding of the changes envisaged in a regime over time, but in relation to territorial priorities. The production of visions is an important participatory process used to engage, inspire and mobilise a wide variety of different social actors, but involves negotiation and struggle. In terms of urban infrastructure, this may involve, *inter alia*, representatives of utilities, municipal government, regulators, developers, business, citizens, 'users', and so on. Visions and the goals they outline provide a reference point through which networks can be built, gaining commitments to 'participate', orientating the actions of potential participants and constituencies, and persuading potential participants of the desirability of transition (see Russell and Williams 2002: 60–61). Although visions are not fixed and will change over time with the variety of social interests that become involved, the key point is that there is an issue of whether visions are initially articulated around narrow coalitions of self-interest – be that from within existing socio-technical regimes or from within narrowly constituted urban governance coalitions – rather than in terms of a more broadly constituted sense of what a purposive urban transition would look like. There is, thus, a crucial issue of who, or which social interests, produce these early visions of the future, and with what expectations.

This is important in view of the fact that the priorities neither of urban governance networks nor of socio-technical regimes are monolithic; they are, as we have seen, constituted by multiple relationalities. Constructing a vision of an urban socio-technical transition encompasses multilevel governance arrangements and socio-technical systems that are often messy, involving multiple actors and institutions across different scales. The construction of coalitions and development of a shared city vision are important but also potentially problematic, given the multiple scales, priorities and social interests involved, the different motivations they have for involvement, and the different financial, knowledge and relational resources they can mobilise. When addressing systemic transitions in urban infrastructures effectively, it is necessary to involve representatives of a wide variety of relevant social interests. This may be more or less problematic, particularly if regimes and systems of infrastructure provision are organised nationally or regionally and ask the question: To what extent are they able to be reorganised and control exerted over the process in relation to the achievement of territorial priorities?

Following from this, the issue of developing a shared vision of urban infrastructure transitions raises two critical questions about who it is that speaks on behalf of 'the city': (1) Who should be involved? (2) And how far should this go? This is important for two reasons in particular. First, it is important in terms of the type of vision that is produced and the range of views that it embodies; and second, it is important that there be engagement at an early stage in a process (see Wilsdon and Willis 2004) of urban socio-technical transition so that there is ownership from and engagement with those whose participation, knowledge and expertise are necessary to translate the vision into practice at a later date. This is not unproblematic. Social interests will have different motivations for, expectations of, and ability to engage in such a process.

Translating visions: intermediary organisation and the capability to act

Visions of purposive urban transitions represent a transformative view of the relationship between cities and socio-technical regimes. Both urban governance networks and socio-technical regimes in and of themselves, by definition, are relatively stable and obdurate. With purposive urban socio-technical transitions, therefore, the aim is to mutually transform both urban governance regimes and socio-technical regimes. This is not straightforward! The production of a vision provides a framework and a direction of travel for a purposive urban socio-technical transition but it says little about how this will be done. A vision in that sense is a necessary but not a sufficient condition of a purposive urban transition. What is required is a sense of how an 'effective' capacity can be coordinated to act on the vision and the process of manifesting that capacity in action, or in other words its capability.

An ad hoc and reactive alignment of social interests will not achieve the priorities encompassed in a vision. Coordinating capacity and mobilising capability require the creation of 'new' intermediary organisational contexts. The creation of intermediaries is necessary to constitute a space outside of the obduracy of both existing urban governance networks and existing socio-technical regimes (Hodson 2008), but that creates a context for the discussion of competing priorities, and in doing so brings social interests from each of them together.

The pressures to reconfigure socio-technical regimes at an urban scale are becoming manifest at a point in history when the governance of these systems is increasingly polycentric, at multiple levels or scales of governance, and control is dispersed and distributed. It is within this context that 'new' forms of governance are emerging, being designed and experimented with to intervene in and seeking to reconfigure regimes at an urban scale. An increasingly central part of these new forms of governance is intermediary organisations that are set up to intervene in a variety of ways in existing systems of producing and consuming resources.

Though intermediaries bear the same generic title, they encompass a wide variety of different organisational priorities and motivations, funding streams and organisational capabilities, predicated on the pursuit of different political priorities aligned with interventions. Though these organisations are frequently different in many respects, including the specificities of their function, they can be characterised in terms of three aspects of their mediating function. First, intermediaries mediate between production and consumption rather than focusing solely on production or consumption issues (see Van Lente *et al.* 2003). Second, they mediate the different priorities and levels of different funders, 'stakeholders', policy interests, social interests, regulators. Third, they also mediate not only between different priorities, in the production of a vision, but also in their 'application'.

Different intermediary organisations fulfil different roles in intervening and seeking to, in some way or another, reconfigure socio-technical regimes. Intermediary organisations can encompass a wide variety of bodies, including government or semi-government energy agencies working at different scales of governance, non-governmental organisations, agencies sponsored by utilities, energy service companies (ESCOs), and so on, that perform functions such as the provision of energy advice and advice centres; consultancy activities; energy audits; project initiation, management and coordination; demonstrations; technology procurement; installation; promotion; advocacy; lobbying, dissemination and awareness raising; the organising of campaigns; education; training and courses; and network building. In doing this, different intermediary organisations function over timescales that can vary from a short-term project or initiative (e.g. lasting six months) to something that is much more long-term and programmatic (e.g. lasting ten years and upwards). Intermediaries can be either project focused or systemic in their orientation, but what we characterise here are systemic intermediaries, given their potential role in purposive urban transitions rather than the more limited role of project intermediaries.

It is important to ask what it is that we can understand about the mediating roles that systemic intermediaries play in intervening in purposive urban socio-technical transitions. It is particularly so in respect of the ways in which intermediaries work at mediating different priorities, and between these priorities and their 'application'. In short, whose interests and priorities shape interventions, and how? The issue we are concerned with is the critical role of systemic intermediaries in developing the organisational capacity necessary to attempt to reconfigure regimes at an urban level. It is crucial to understand the negotiation of whose priorities it is that shapes these responses. What is the balance between social interests from 'outside' of a city, whether that is national government priorities, regulators, utilities, etc., and territorial priorities? These different social interests each bring not only different expectations of a transition but also forms of knowledge, expertise and understandings.

The basis upon which these social interests, their expectations and forms of knowledge are organised is crucial to underpinning the development of active capacity and necessary capability to translate a vision into social and material action. In our previous work on intermediaries and through analysis on European intermediary practices (Hodson and Marvin 2009b, c), we have identified seven issues (see Table 5.1) that are particularly important in constituting capacity and capability. Through addressing these seven critical issues, we have considered the necessary organisational context for how intermediaries can intervene in purposive urban socio-technical transitions actively and effectively. Inevitably these issues require further development through translation, practice and refinement. They also require an understanding of how we would know whether interventions were 'effective' and 'successful'.

Table 5.1 A framework for active and configurational intermediation

1 Financial issues	Develop a context of broad-based and stable sources of funding that offers the potential for financial independence. This creates the conditions where intermediary priorities are not largely dictated by the reactive chasing of funding and the priorities of different funders. This is important in creating stability in relation to a series of further issues – see below.
2 Staffing	Security of funding underpins the security of core employee positions. It creates stability and a backdrop where staff training and skills programmes can be developed. Stability means that resources are available so that staff and employees can be incentivised, feel rewarded and not subject to the whims of short-term funding – forming the basis for an organisational commitment to the careers of employees.
3 Organisational structures and cultures	This is particularly important where many intermediaries work with a small core with a broad network of partners, and a stability of organisational resources and commitment provides the basis for a shared culture and clarity around different organisational positions. Small capacities require intermediaries to effectively 'plug in' to networks of partners to enhance capacity – but from a shared organisational view. This very dynamic set of circumstances means that intermediaries must develop as effective learning cultures and develop the ability to adapt to changing pressures and new issues through systemic and strategic thinking and long-term funding.
4 Knowledge base	The adaptability and learning required by intermediaries means that they must constantly work at developing and redeveloping the knowledge base to which they have access – where a wide variety of technical, policy and local forms of knowledge need to be constantly negotiated and effectively integrated.
5 Communications	This requires the alignment of different sets of social interests and their priorities, and the creation of communications forums to make this possible. It means that intermediaries need to develop a local presence and good local networks through proximity and face-to-face communications. Intermediaries also need to develop effective relationships and resources, beyond local networks, with national policy makers.
6 Credibility	Intermediaries need to think carefully about how they represent what they do to others. This is important in communicating credibility and building trust with a variety of partners, who in other aspects of their work and business may have competing priorities. Symbolic visibility in the local and national media is important, as is symbolic exemplification through demonstration and showcasing. This is part of the positioning of the intermediary as distinctive, as 'first mover' and 'the people to turn to'.
7 Influence	The previous six issues are important in embedding the intermediary within a local context and facilitating the development of the resources, relationships, forms of knowledge and communications and, thus, visibility, to be able to have credible influence. The intermediary needs to develop a shared organisational view of how it would know whether it was influential beyond narrow metrics of funders.

Urban transitions: how would we know?

Understanding urban socio-technical transitions through the lens of the interventions of intermediary organisations requires us to ask: How would we know whether interventions had been 'effective' and 'successful'? There are two particular issues that need to be addressed here in constructing an initial analytical framework that can tackle this issue.

The first is the extent to which the aims, objectives and aspirations of the vision are achieved over time – an 'outcome' indicator of 'effectiveness' and 'successfulness'. Thinking about (in)effective and (un)successful outcomes acknowledges the importance of a focus on the intermediary organisation as an agent of change. In particular, it requires addressing the degree of resonance or dissonance between the initial vision of urban socio-technical transition and its achievement over time, in respect of aims, objectives, timings, material and social change and so on. The extent of the similarities and/or gaps – between the aims and objectives proposed in the vision and their outcomes – informs an understanding of the degree of outcome success. Strong resonances between the objectives, application, timescales and budgets would inform a high degree of outcome success, while gaps between the objectives outlined in the vision and their realisation would inform a high degree of unsuccessfulness in outcomes. Obviously, the achievement of some objectives would indicate a degree of success somewhere in between. The emphasis on outcome success allows us to retain a focus on the vision and its objectives in urban socio-technical transitions. However, it tells us little about the processes and politics through which the vision achieves or fails to achieve 'acceptance' among a wide variety of stakeholders and translation into materiality.

The second issue is the extent to which these aims, objectives and aspirations are embedded in social practices – that is, a more processual and contextual view of 'effectiveness' and 'successfulness'. Our starting point here is again with the intermediary and with the objectives of a vision. Either implicitly or explicitly captured within visions is a sense of whom intermediaries 'need' to engage to translate the vision. This may be broad-ranging or narrow in terms of the types of social interests – for example, funders, planners, users, residents, technology suppliers, local authorities, national governments and so on. – intermediaries anticipate they will need to engage with. What is of particular interest is that having engaged with the different social interests in the process, intermediaries may still be confronted with difficult issues and problems in physically dense and congested urban cores. This might include, for example, controversies such as where a technology development is located, difficulties with funding streams, lack of political support and so on. How these issues are addressed and who subsequently becomes involved and with what expectations then become critical to the process. This is important, as it broadens the constituency of the process of urban socio-technical transitions. A controversial location for technology development, for example, may involve technology developers engaging with local residents; funding difficulties may require dialogue with different funding bodies; a lack of political support may involve discussions with political interests at different levels. Each of these social interests potentially brings different sets of expectations to the process of urban socio-technical transitions.

The intermediary coordination of these different social interests (or capacity) is the key signifier of process success and 'acceptance'. Coordination may occur between different social interests through a variety of methods and media. Addressing a controversial location for technology development may be through public meetings, via public information leaflets, through planning processes and so on. Likewise, a funding problem may be addressed through face-to-face meetings and bids for funding. A lack of political support could involve intermediaries trying to build relationships through lobbying politicians, through a media offensive and so on.

A 'totally successful' process would have a fully coordinated constituency at the 'end point' of an urban socio-technical transition. Those whom intermediaries need to engage to realise the vision would have been successfully enrolled. Any issues, controversies or problems that arose would subsequently have been addressed through involvement of the 'necessary' social interests and the 'relevant' resources and be coordinated – through various methods – with the initial objectives of the vision. A 'totally unsuccessful' process would have failed to engage with those needed to realise a vision. Any further issues, controversies or problems that arose would not have subsequently been addressed. The involvement of the 'necessary' social interests and the 'relevant' resources would not have been sought and there would, therefore, be no coordination with the initial objectives of the vision.

Of course, the likelihood is that many intermediary interventions over time would be located on a continuum somewhere between the two points. The development of these ways of thinking about outcome and process success is not an end point in itself but provides the basis for understanding and thinking through intermediary interventions in urban socio-technical transitions and will be adapted and reworked over time as they are applied to the context of different attempts at urban socio-technical transitions.

Conclusions

This chapter asked two questions: Can cities shape socio-technical transitions? And how would we know if they were doing so? We return to those questions here. Asking these questions arose from our interest in some of the world's most powerful cities' attempts to purposively reconfigure socio-technical regimes at the scale of the city. In the chapter, we engaged with the MLP on socio-technical transitions to address the ways in which it does and does not provide a conceptual framework through which to understand these trends and developments. This enabled us to ask questions about how we understand questions of geography and difference between socio-technical regimes. This was particularly important in a context where, increasingly, urban territorial priorities appear to be informing attempts by urban political elites to gain degrees of control over the organisation and functioning or energy, water, waste and transport infrastructures – often where they have limited or no involvement in socio-technical regimes. We developed a framework for understanding a purposive urban energy transition, how shared understanding of this could be developed by various territorial and regime interests, the intermediary organisational context for how interventions to translate these visions would be constituted and how understandings of whether these interventions had been successful or not would be constituted.

In doing this, we recognise that we are contributing to a nascent debate and that our framework provides a basis for intervention but is one that will be informed and adapted by future empirical investigation. Consequently, the chapter has contributed to three issues.

First, our analysis highlights a key challenge for transitions researchers related to spatial contexts: Where is the 'where' of transitions approaches? The interpretative flexibility of many transitions concepts is what in many ways makes transitions approaches resonate with such a wide variety of researchers and practitioner interests. Yet when one is utilising these concepts in relation to particular problems, contexts or scales, these terms need to be clarified. Central to this chapter has been both the questioning of the extent to which key transition concepts are applicable to urban contexts and the attempt to empirically interrogate key transitions concepts in a particular urban context. In future, much more of this conceptual-empirical engagement around urban transitions needs to be undertaken.

Second, conceptually this not only forces us to clarify where key conceptual units are but in doing so links where conceptual units, like regimes, are and how they may be rescaled in transition from a national to an urban system of provision. This brings together spatial with temporal issues. Methodologically, attempts to research the future are notoriously hazardous. How do you research 2020 or 2050? What we have contributed to is an engagement with this problematic through the development of a framework for researching claims about the future of the city (to 2025, 2050 and 2100) and how the reorganisation of its socio-technical infrastructures can be undertaken. Much more of this contextually sensitive research in various urban contexts is required, as is a comparative understanding of the similarities and differences between urban contexts.

Finally, what we have been examining in this chapter is the extent to which socio-technical systems and their transition can be governed and configured at an urban scale. We have demonstrated that there is a need for an effective coordination of capacity and capability to initiate and attempt to enact systemic transitions. This poses two further issues for research. The first of these relates to the translation over time of the visions of systemic socio-technical transitions and the necessity of undertaking ongoing research into attempts to enact transitions and the political processes and participation involved in doing so; the second is the issue of what happens in cities that do not have the resources and capacities to mobilise that world cities have. Further research should engage with transitions in cities other than premium world cities and examine what transitions look like in ordinary cities, and cities of the global South.

References

Bache, I. and Flinders, M. (eds) (2004) *Multi-level Governance*, Oxford: Oxford University Press.

Berkhout, F., Smith, A. and Stirling, A. (2003) 'Socio-technological regimes and transition contexts', Working Chapter series, SPRU, University of Sussex, Brighton.

Brenner, N. (2004) *New State Spaces: Urban Governance and the Rescaling of Statehood*, New York: Oxford University Press.

Brenner, N. and Keil, R. (2006) *The Global Cities Reader*, New York: Routledge.

Dalby, S. (2007) 'Anthropocene geopolitics: Globalisation, empire, environment and critique', *Geography Compass* 1 (1): 103–118.

Elzen, B., Geels, F. W. and Green, K. (eds) (2004) *System Innovation and the Transition to Sustainability: Theory, Evidence and Policy*, Cheltenham, UK: Edward Elgar.

Geels, F. W. (2002a) 'Technological transitions as evolutionary reconfiguration processes: A multi-level perspective and a case study', *Research Policy* 31: 1257–1274.

Geels, F. W. (2002b) Towards sociotechnical scenarios and reflexive anticipation: Using patterns and regularities in technology dynamics', in K. Sørensen and R. Williams (eds) *Shaping Technology, Guiding Policy: Concepts, Spaces and Tools*, Cheltenham, UK: Edward Elgar.

Geels, F. W. (2004) 'From sectoral systems of innovation to socio-technical systems. Insights about dynamics and change from sociology and institutional theory', *Research Policy* 33: 897–920.

Geels, F. W. (2005) 'Co-evolution of technology and society: The transition in water supply and personal hygiene in the Netherlands (1850–1930) – a case study in multi-level perspective', *Technology in Society* 27 (3): 363–397.

Geels, F. W. and Schot, J. (2007) 'Typology of socio-technical transition pathways', *Research Policy* 36 (3): 399–417.

Graham, S. and Marvin, S. (2001) *Splintering Urbanism*, London: Routledge.

Green, K. and Foster, C. (2005) 'Give peas a chance: Transformations in food consumption and production systems', *Technological Forecasting and Social Change* 72: 663–679.

Harvey, D. (1989) 'From managerialism to entrepreneurialism: The transformation in urban governance in late capitalism', *Geografiska Annaler, Series B* 7: 13–17.

Hodson, M. (2008) 'Old industrial regions, technology and innovation: Tensions of obduracy and transformation', *Environment and Planning A* 40 (5): 1057–1075.

Hodson, M. and Marvin, S. (2009a) ' "Urban ecological security": A new urban paradigm?', *International Journal of Urban and Regional Research* 33 (1): 193–215.

Hodson, M. and Marvin, S. (2009b) 'Cities mediating technological transitions: Understanding visions, intermediation and consequences', *Technology Analysis and Strategic Management* 21 (4): 515–534.

Hodson, M. and Marvin, S. (2009c) *Identification of Intermediary Practices across Countries for Assessing Piloting*, Report for Changing Behaviour Project, Seventh Framework Programme, European Commission.

Hodson, M. and Marvin, S. (2010) *World Cities and Climate Change*, Maidenhead, UK: McGraw-Hill.

Hoogma, R., Kemp, R., Schot, J. and Truffer, B. (2002) *Experimenting for Sustainable Transport: The Approach of Strategic Niche Management*, London: Spon Press.

IEA (2008) *World Energy Outlook 2008*, Paris: International Energy Agency.

Jessop, B. (1997) 'The entrepreneurial city: Re-imaging localities, redesigning economic governance, or restructuring capital?', in N. Jewson and S. MacGregor (eds)

Transforming Cities: New Spatial Divisions and Social Transformation, London: Routledge.

Kemp, R. and Loorbach, D. (2005) 'Dutch policies to manage the transition to sustainable energy', in F. Beckenbach, U. Hampicke and C. Leipert (eds) *Jahrbuch ökologische Ökonomik: Innovationen und Transformation*, Band 4, Marburg, Germany: Metropolis.

Luke, T. W. (2003) 'Codes, collectivities, and commodities: Rethinking global cities as megalogistical spaces', in L. Krause and P. Petro (eds) *Global Cities: Cinema, Architecture, and Urbanism in a Digital Age*, Piscataway, NJ: Rutgers University Press.

Monaghan, A., Hodson, M. and Marvin, S. (2009) 'The role of cities and regions in socio-technical transitions: Towards a typology', Working Chapter, SURF.

Rotmans, J., Kemp, R. and van Asselt, M. (2001) 'More evolution than revolution', *Foresight* 3 (1): 1–17.

Russell, S. and Williams, R. (2002) 'Social shaping of technology: Frameworks, findings and implications for policy', in K. H. Sørensen, K. and R. Williams (eds) *Shaping Technology, Guiding Policy: Concepts, Spaces and Tools*, Cheltenham, UK: Edward Elgar.

Sassen, S. (2001) *The Global City*, Princeton, NJ: Princeton University Press.

Smith, A., Stirling, A. and Berkhout, F. (2005) 'The governance of sustainable socio-technical transitions', *Research Policy* 34 (10): 1491–1510.

United Nations (2006) *World Urbanization Prospects: The 2005 Revision*, Department of Economic and Social Affairs, Population Division, United Nations, New York.

van der Brugge, R. and Rotmans, J. (2007) 'Towards transition management of European water resources', *Water Resources Management* 21 (1): 249–267.

Van Lente, H., Hekkert, M., Smits, R. and van Waveren, B. (2003) 'Roles of systemic intermediaries in transition processes', *International Journal of Innovation Management* 7 (3): 247–279.

Verbong, G. P. J. and Geels, F. W. (2007) 'The ongoing energy transition: Lessons from a socio-technical, multi-level analysis of the Dutch electricity system (1960–2004)', *Energy Policy* 35 (2): 1025–1037.

Voβ, J.-P., Bauknecht, D. and Kemp, R. (eds) (2006) *Reflexive Governance for Sustainable Development*, Cheltenham, UK: Edward Elgar.

Wilsdon, J. and Willis, R. (2004) *See-Through Science*, London: Demos.

Part II

Urban transitions in practice

6 Urban energy transitions in Chinese cities

Shobhakar Dhakal

Introduction

In 2005, China's urbanization rate was 40 per cent, representing 531 million urban dwellers, which is 17 per cent of the world's urban population (UN 2007). Increasing urbanization is a national policy priority in China, where urbanization has been perceived by decision makers as an element necessary to economic and industrial growth. Accordingly, China's Eleventh Five-Year Plan aims to increase the urbanization rate to 47 per cent by 2010 (Raufer 2007). UN projections show that urbanization in China will grow rapidly, reaching 73 per cent, or 1.02 billion urban dwellers, in 2050. However, increased urbanization demands greater energy use in a rapidly developing economy, owing to rising household incomes and the continuing concentration of energy-consuming sectors into urban areas. Rising incomes make urban dwellers' lifestyles more energy-intensive, and the new urban migrants demand greater energy per capita than they did in their rural settlements. Consequently, urban areas will play a greater role than at present in shaping China's energy demand and CO_2 emissions. In effect, in the past three decades government policies in China have created an 'energy-intensive' and 'high carbon' urban transition through their efforts to increase urbanization and economic growth. However, energy use and carbon emissions per capita are still small in China and, as this chapter demonstrates, efforts to develop 'low carbon' urban futures are also emerging.

Because of its high demand for energy, China is already aggressively engaged in securing energy resources internationally and has already passed the United States as the largest emitter of CO_2 (Global Carbon Project 2008). Thus, energy security and climate change mitigation have emerged as key policy priorities in China.

Additionally, pollution is a critical concern in China's urban areas. Since coal and oil dominate (urban) energy systems, they are intricately linked to the increases in particulate emissions, acid rain and automobile pollution such as NO^x and ozone. Sixteen Chinese cities are listed among the world's twenty most polluted cities, and this has alarmed policy makers (The Economist 2004). Urban energy systems and the urban infrastructure are at the centre of the necessary countermeasures to tackle issues of security, climate change and pollution and could make positive impacts that might last for several decades (Raufer 2007). In China, however, energy concerns have historically played a minor role in urban planning (ibid.). China's Eleventh Five-Year Plan recognizes the importance of energy for economic growth but fails to sufficiently recognize the opportunities that exist to improve the efficiency of the urban system as a whole (Wen 2005).

In this context, the chapter is structured in five sections. The first section examines the urban contribution to China's energy uses and thus highlights their importance in determining the energy and carbon profiles of the nation. This helps illustrate how important urban areas in China are for key ongoing national concerns such as improving energy security, mitigating climate change and substantially reducing the energy intensity of Chinese economy. The second section examines the internal dynamics of urban areas at the meso scale by estimating and analysing the energy usage and CO_2 emissions of the highly urbanized and economically most important thirty-five cities, as defined by China's National Plan, demonstrating the extent to which a 'high carbon' urban energy transition has been taking place. The third section examines this in more detail, providing a review of the changes in urban energy uses and CO_2 emissions and their drivers at the micro scale through a detailed analysis of Beijing, Shanghai and Tianjin. The fourth section focuses on a recent study of Shanghai and looks in detail at the carbon emissions associated with two different development scenarios. In conclusion, the future policy implications of these urban energy transitions are considered.

China's urban energy consumption

Detailed information on energy usage is easily available only for a few large cities in China. The greatest contribution to China's urban population comes from the smaller urban agglomerations (agglomerations of population less than 1 million account for 57 per cent of total urban population in 2005) whose information base is poor, and the smallest contribution comes from the larger urban agglomerations (especially those above 5 million, which account for 12 per cent of total urban population), whose information base is relatively better. Consequently, this chapter uses an aggregated methodology for estimating urban energy uses in China. This is based on the energy intensity of economic activities, which is the energy consumption per unit gross regional product (GRP), as a key indicator on the basis of which urban energy usage in China in 2006 is estimated.[1] The results of these estimations provide important insights into the urban contribution of energy use (Table 6.1).

The urban contribution to China's total energy use is enormous – 84 per cent of energy consumption in 2006. The gap between the urban and rural populations for energy use is massive: the ratio between urban and rural energy usage per capita is 6.8, and that between urban and national usage is 1.9. However, if we consider fuels with lower calorific values such as biomass and other forms of renewable energy, the urban contribution is expected to be slightly lower, owing to the large biomass usage in the rural areas. Statistics show that biomass and other

Table 6.1 China's urban energy consumption (estimated), 2006

		Unit
Total urban energy consumption	2,071	million tons of SCE (standard coal equivalent)
Urban energy consumption per capita	3.59	tons of SCE/person
Rural energy consumption per capita	0.53	tons of SCE/person
Total energy consumption per capita	1.87	tons of SCE/person
Urban to national per capita energy ratio	1.92	
Urban energy's share in total	84.07%	Based on national energy as 2,032 and 2,463 million tons of SCE respectively

Source: Dhakal (2009).

renewable sources of energy contributed 10.6 per cent of the total energy uses in China in 2006 (IEA 2007). Consequently, if we take these IEA numbers for informal energy use and further assume that the urban areas use only formal energy sources, then the urban contribution to the total energy consumption of China would be 75 per cent in 2006.

Energy use in China's thirty-five key cities

In this section, I analyse and illustrate the internal dynamics of urban areas through the differences and commonalities in energy usage between cities that are highly urbanized and explicitly designated in China's National Plan as important cities for economic development. The National Plan of China has designated thirty-five cities as key cities; they include provincial capitals, provincial cities and other economically important cities as the most important cities in China for economic growth and infrastructure development.[2] Effectively, these are the places through which China's 'urban transition' is being forged. Against the backdrop of rapid urbanization and the growing energy intensity of urban life, it will be these cities that take the lead in the implementation of key policy measures in urban energy efficiency and climate change mitigation. For this analysis, I first estimate the contributions of these thirty-five cities and then show the different energy transition pathways that they exhibit. Such a comparison, even though it is derived from a modest methodology, is the first published comparison. In order to estimate the total energy consumption from these cities, the average carbon and energy intensity (energy/carbon per unit GRP) of the province to which the city belongs is used as a proxy for the city's energy and carbon intensity.

The analysis, shown in Table 6.2, shows that these cities represent less than one-fifth of the population but produce a large share of the nation's GDP. Collectively, they consume 40 per cent of the total commercial energy of the nation and emit CO_2 at similar levels. The wide disparity between these cities and the nation in GDP per capita, energy consumption per capita and CO_2 emissions per capita

Table 6.2 Key indicators and estimated energy and CO_2 for the key thirty-five cities of China, 2006

	China	*Frontrunner cities*	*Cities' contribution*
Total population, million	1, 314	237	18%
GRP (market price), billion US$	2,719	1,109	41%
Total commercial energy consumption, million TJ	65.7	26.2	40%
Commercial energy consumption per capita, MJ/person (registered permanent population)	50,000	110,771	2.2 times more
GDP/GRP per capita, US$/person (registered permanent population)	2,068	4,681	2.3 times more
CO_2 emissions (commercial energy-related), million tons	5,645	2,259	40%
CO_2 emissions per capita (commercial energy-related), tons/person	4.30	9.54	2.2 times more

Sources: Calculated from base data of CSY (2007) and CESY (2007) by author. See Dhakal (2009) for details.

shows that their influence in shaping the national energy and carbon profile is disproportionate compared to their population.

Apart from the differences between these cities and the nation, there are large differences within these cities, as demonstrated in Figure 6.1. Among these cities, Shenzhen stands out with extraordinarily high GRP per capita (US$38,000 per person). Shenzhen is a gateway to China from Hong Kong and is a centre for financial markets and other manufacturing establishments. Figure 6.1 shows three clear pathways in China's thirty-five key cities. The first is the low energy consumption and high economic output path adopted by cities like Fuzhou, Nanjing, Ningbo, Beijing, Guangzhou, Shanghai and Xiamen; the second is the high energy consumption and low economic output path adopted by cities such as Xining, Yinchuan, Guiyang, Urumqi, Taiyuan, Hohhot and others; and the third lies between these two. Cities in the first group are largely situated in the eastern part of the country, close to the coast with a relatively warmer climate, and have a strong presence of service industries, while cities in the second group tend to be more inland – in central and western China, with energy-intensive industries and relatively cooler climates. However, the criteria and indicators for comparing cities' energy performance are often not straightforward, for several reasons. The role of a city as a manufacturing hub, its prevailing climate, its income and many other factors affect its energy use, but local policies may allow only limited or no control over many of the indicators used for evaluating the city's energy or carbon performance. At the same time, decoupling the role of manufacturing industry could be difficult because urbanization and industrialization are closely linked in many cases.

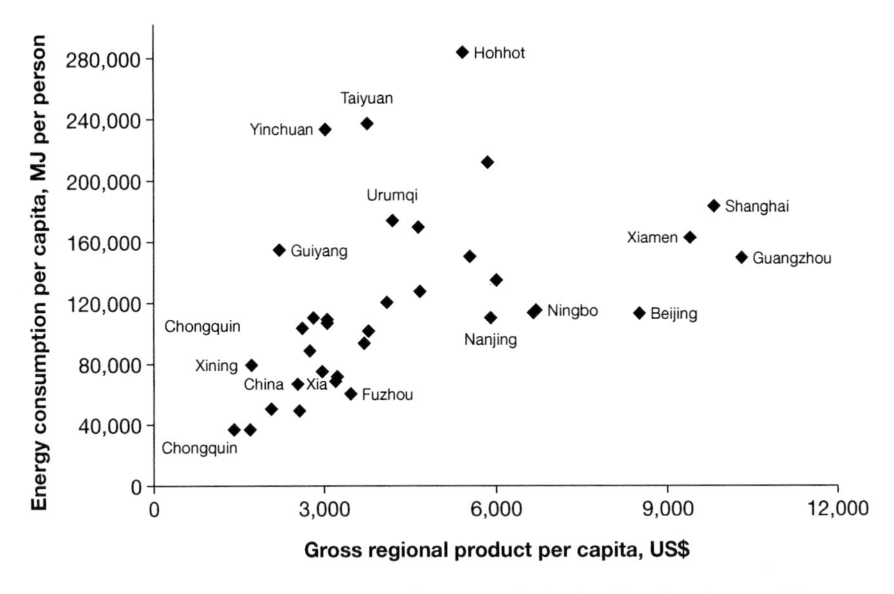

Figure 6.1 Income and energy consumption per capita in the thirty-five largest Chinese cities, 2006.

Source: Estimated by the author, based on CSY (2007) and CESY (2007).

Energy use and CO_2 emission changes in Beijing, Shanghai and Tianjin

In this section, I carry out a more detailed historical analysis of energy uses and carbon emissions from three highly urbanized and trendsetting cities, namely Beijing, Shanghai and Tianjin, in order to explain the drivers of urban energy system transitions, where transitions are understood in historical terms, and the resulting urban energy usage and CO_2 emission transitions. The historical changes in urban energy use and CO_2 emissions that are taking place in these key cities help illustrate the speed and magnitude of the urban energy and CO_2 changes taking place in Chinese cities. These cities are the three most prominent megacities whose populations exceed 10 million, and they have provincial status in China's political and administrative system. Beijing, Shanghai and Tianjin have been provincial cities for many decades and are highly urbanized. Table 6.3 outlines the basic indicators for these cities.

The analysis shows a rapid growth in energy uses and CO_2 emissions from each of these cities, which is seen to have accelerated from the early 2000s. In Shanghai, emissions have increased by 5 times between 1985 and 2006, in contrast to 2.6 times for Beijing and 2.8 times for Tianjin in the same period. Even in the period 2000–2006 alone, these cities increased their emissions by a factor of 1.5–1.7. Until 1990, the emissions difference between Beijing and Shanghai was nominal,

Table 6.3 Basic indicators for Beijing, Shanghai and Tianjin, 2006

	Beijing	*Shanghai*	*Tianjin*
Area, sq. km	16,410	6,340	11,920
Resident population, million	15.81	18.15	10.75
Registered population, million	11.98	13.68	9.49
Urban share in resident population (%)	84%	89%	76%
Gross regional product, billion US$	98.7	130.0	54.7
Total energy use, thousand TJ[a]	1,332	2,480	1,271
Total energy-related CO_2 emissions, million tons[a]	142.10	228.74	117.61

Sources: Statistical Yearbook of each of four cities and CSY (2007).

Note

Exchange rates are from the database of the International Monetary Fund.

a Author's own calculation based on the energy balance table for each city from CESY (2007).

with Shanghai slightly higher (in 1985, Beijing's emissions were higher than Shanghai's), but it started to widen after China's economic growth accelerated. By 2006, the gap in emissions among the three cities was large. In 1995, the energy use and CO_2 emissions per capita (registered population) of Beijing, Tianjin and Shanghai were all close to 2 tons of oil equivalent and 8 tons respectively; however, by 2006 the CO_2 emissions per capita for Beijing and Shanghai would be 9 tons and 12.6 tons respectively.

The sectoral analysis shows that the industrial sector (see Figure 6.2) has historically dominated CO_2 emissions in these cities. In particular, Beijing and Shanghai have gone through rapid transformations in the period 1985–2006, characterized by the rapidly declining share of the industrial sector in total CO_2 emissions – 65 to 43 per cent for Beijing and 75 to 64 per cent for Shanghai – and the rising share for the commercial and transportation sectors. This is an indicator of rapid urban transition. This transformation has been faster in Beijing than in Shanghai and Tianjin. The residential sector too has grown tremendously in energy use and carbon emissions, but because of the rapid growth and expansion of other energy-consuming sectors the share of energy used and carbon emissions by residential sectors have remained more or less unchanged over the past two decades for Beijing, Shanghai and Tianjin. In the case of the transportation sector,[3] the shares in 2006 are relatively smaller – 7 per cent for Tianjin and 16 per cent for Beijing – but their growth rates have been very high, owing to the rising car ownership rates in these cities. Despite strong control over vehicle ownership in Shanghai,[4] the CO_2 emissions from the transportation sector have increased by eight times in the period 1985–2006. Beijing registered close to a sevenfold increase in the same period. In the past decade alone, CO_2 emissions in Tianjin have increased by a factor of almost 3.5.

With respect to fuel usage, a rapid change in CO_2 emissions is taking place as well. First, a rapidly declining trend for the share of direct coal burning in both energy consumption and CO_2 emissions is evident. In Beijing, Shanghai and

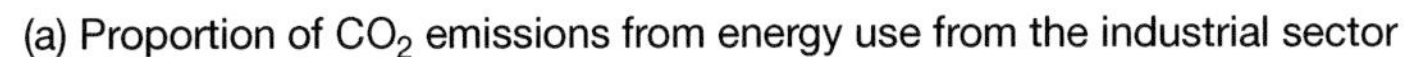

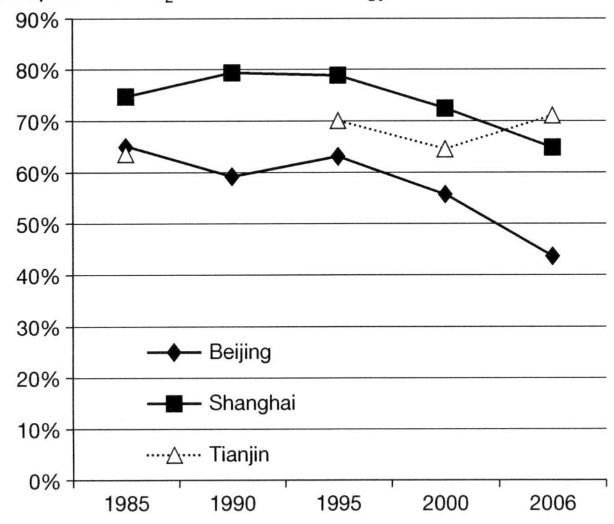

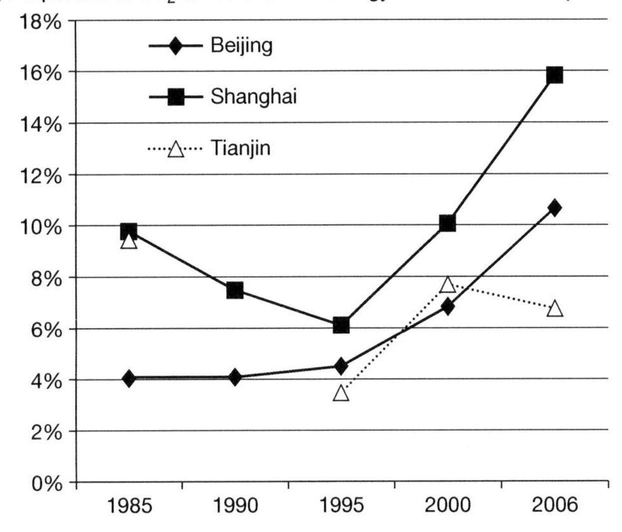

Figure 6.2 Share of sectors in energy-related CO_2 emissions in cities.

Source: Dhakal (2009).

Tianjin, coal's share of CO_2 emissions declined from 58 to 26 per cent, 51 to 18 per cent, and 61 to 33 per cent, respectively, between 1985 and 2006, thus showing a rapid transition. Second, the share of electricity and oil in CO_2 emissions and energy consumption is rising, which is compensating for the declining share of coal burning. For Beijing, CO_2 emissions from electricity use increased by more than fivefold between 1985 and 2006, and its share doubled from 20 to 40 per cent over the same period. Shanghai is similar to Beijing, and in Tianjin CO_2 growth from electricity use is relatively slower. The rise in electricity usage is more noticeable than that for oil in all cities except Shanghai, where both electricity and oil are main features of the energy system. In Shanghai, oil contributed 50 per cent of the final energy consumption in 2006, a rate much higher than for the other cities. Third, the role of natural gas in the cities' energy systems has increased, especially since 1995. Natural gas, a cleaner source of energy, made up 10, 3 and 4 per cent, respectively, of the final energy consumption of Beijing, Shanghai and Tianjin in 2006, but the share that natural gas contributes to the total CO_2 emissions is close to half of the share that natural gas contributes to total energy consumption. Beijing, in particular, rapidly developed an infrastructure for natural gas in a short period of time; until 1997, natural gas represented only 0.8 per cent of the total final energy consumption, a proportion that had increased to 10 per cent by 2006.

Analyses of the relative importance of the driving factors of CO_2 emissions from energy uses, using the factor decomposition method (see Dhakal 2009 for details), show that economic growth and energy intensity have dominant effects on the changes in CO_2 emissions in all cities. In this method, the changes in CO_2 emissions can be attributed to the changes in four factors separately: the emission per unit energy use (fuel type and quality), the energy intensity of gross regional product (energy efficiency and economic structure change), GRP per capita (income) and population. The analysis is able to show the relative importance of each of the factors to the changes in CO_2 emissions, and helps to explain the observed changes in CO_2 emissions. It shows that economic growth drove CO_2 emissions, but improvement in energy intensity helped to dampen the CO_2 growth. The forces of these two factors from 1995 to 2000, in particular, were very intense, as this period was marked by slowdowns in emission growth in the cities and for China as a whole. In addition, the impacts of fuel shifts to the changes in CO_2 emissions are nominal in these cities, owing to the dominance of coal and oil. The shift to cleaner fuels is not significant and thus has not contributed to the dampening of CO_2 growth, despite the perceived impression that the role of natural gas is rapidly expanding in Chinese cities' energy systems. Moreover, the analyses show that the carbon emitted per unit of energy consumption is slightly increasing in the cities. The population effect has contributed to increased CO_2 emissions consistently over the years. When the floating population is considered in the analyses, the impact of population to the changes in CO_2 emissions was found to be much higher in Shanghai.

Based on the earlier estimations, the cities' historical changes in terms of carbon intensity of economic activities and CO_2 emissions per capita are shown in Figure 6.3. This figure illustrates several key observations regarding the urban energy and CO_2 transitions in the cities. It shows that the CO_2 emissions per capita in these cities have increased relentlessly over the past two decades with the rapid economic growth. The carbon intensity of the economic activities in these cities peaked in the 1990s, when economic growth accelerated because of the

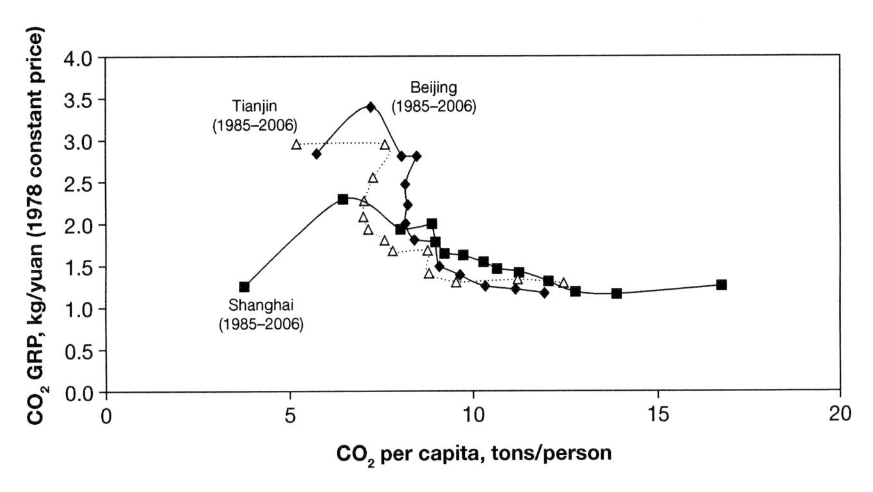

Figure 6.3 Carbon-economy performance.

Source: Dhakal (2009).

contribution of the energy-intensive manufacturing sector. Efforts directed towards changing the economic structure, modernizing and privatizing state-owned enterprises, and improving the energy efficiency of industries were just starting to take shape. Therefore, the energy intensity of economic output of cities increased and thus contributed to increasing emissions. However, throughout the 1990s the energy efficiency of industries in these cities improved rapidly, owing to the restructuring of state-owned enterprises, the fact that better technologies started to be used, the fact that less energy-intensive sectors such as the service sector started to grow rapidly in cities, and the elimination of the use of low-grade fuels such as coal in residential and commercial boilers in households and businesses. These changes resulted into the reductions in the carbon intensity of economic growth, serving to dampen the growth of CO_2 emissions.

Since the early 2000s, the improvement in carbon intensity of economic activities has slowed in all these cities, especially in Shanghai and Tianjin in recent years (Figure 6.3). This may be because of slow growth in the share of the tertiary sector for Beijing in the period 2000–2006 as compared to earlier decades, and the negative growth in shares for Shanghai and Tianjin.

Collectively, for better urban energy and carbon transitions, these analyses of drivers show that the quality, not merely the quantity, of economic growth in cities matters (diversification of value-added sources); cities can shape China's energy transition through optimizing urban systems in terms of public transport infrastructure, buildings, urban planning and providing cities as a test-bed for the implementation of innovative technology and policy ideas. Rapidly rising energy use and CO_2 emissions along with growing service, transport and buildings sectors shows that urban dimensions are causing rapid energy transitions. Cities are the places where energy transitions are happening because cities are centres of economic growth and industrialization. Thus, there is a need to use a large amount of energy in industry. The role of clean energy is determined by industry's proximity to clean energy infrastructure such as a supply of natural gas, and opportunities to tap large-scale efficiency gains in infrastructure. At the same time, cities themselves are also creating particular forms of energy transitions, given their demographic changes, lifestyle changes spurred by economic growth (car dependency and greater energy use in buildings) and the growing levels of consumption.

Carbon emission scenarios in Shanghai

In this section, we consider the specific case of Shanghai, which has been subject to more detailed scenario development work (see Li *et al.* 2009 for details), which estimated that total territorial CO_2 emissions in 2006 in Shanghai were 184 million tons (excluding CO_2 emitted upstream as a result of imported electricity).[5] Under the business-as-usual (BAU) scenario, the corresponding figures will increase to 290 and 630 million tons in 2010 and 2020, respectively, as shown in Figure 6.4. Under this BAU scenario, if Shanghai fails to act in relation to energy conservation and CO_2 emission reduction, the future rise in energy use and CO_2 emissions will be huge: 3.6 times for energy and 3.4 times for CO_2 in 2020 as compared with 2006.

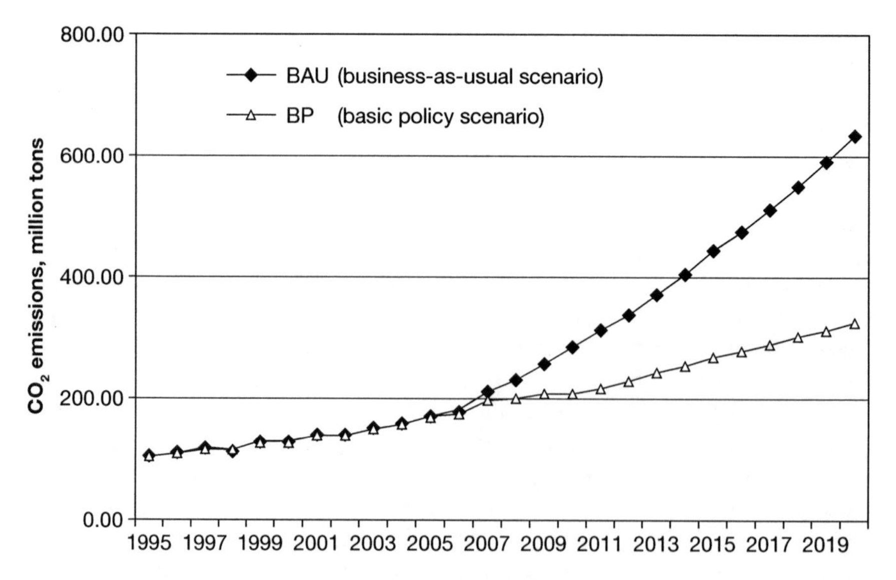

Figure 6.4 Trends in CO_2 emissions for two scenarios.

Source: Li *et al.* (2009).

However, while, historically, urban energy transitions have propelled Shanghai and other cities towards a 'high carbon' future, as discussed above, China's Eleventh Five-Year National Plan has the explicit aim of reducing the energy intensity of gross domestic product by 20 per cent between 2005 and 2010. Accordingly, at national, provincial and city levels China has implemented targeted policies of industrial restructuring, energy efficiency improvements and others to meet this target, and such targets have even appeared in the performance evaluations of city and provincial officials. Shanghai has plans and policies for energy in place until 2010 that may affect CO_2 emissions positively. Improving energy efficiency is one of the top policy agendas of Shanghai explicitly illustrated in the Five-Year Plan (2005–2010) and accompanying sectoral documents of Shanghai's government.

The target of reducing energy intensity by 20 per cent, passed down to cities from the national goal, requires considerable change in Shanghai. The objectives of these plans are economic structural adjustment (more commercial and service sectors), cleaner fuel structure, less reliance on energy-intensive industries, and an increase in energy efficiency at all levels. The basic policy (BP) scenario (Figure 6.4) considers major policies that are already in place and extends them beyond 2010 until 2020. Shanghai's major policies considered here essentially fall under five broad categories: maintaining high economic growth rates of close to 10 per cent; maintaining population at 20 million by 2020; increasing the GRP contribution of the service sector through economic structural adjustment; reducing the share of coal in primary energy consumption from 53 per cent in 2005 to 46 per cent in 2010; and reducing the energy intensity of the industrial sector by 35 per cent, that of the tertiary sector by 15 per cent and energy use in buildings by 15 per cent between 2005 and 2010. Under the BP scenario, CO_2 emissions in 2010 and 2020 will be 210 and 330 million tons, respectively. These amounts are 27 per cent and 49 per cent lower than those of the BAU scenario, as shown in Figure 6.4. These figures reinforce the need for such plans and policies in order to reduce energy consumption and CO_2 emissions in large quantities. Even under the BP scenario, it is not expected that energy use and CO_2 per capita will decrease from 2005 to 2020 because Shanghai will be critical to supporting national economic growth. Consequently, the best that Shanghai can currently do is to slow the growth of CO_2 emissions.

As shown above, the high economic growth rate and associated industrial and urban activities are the major drivers for the tremendous rise in energy use and CO_2 emissions in Shanghai in the past decade in all sectors. Despite a partial structural shift towards natural gas, the energy system of Shanghai is still heavily dominated by coal. Unless carbon capture and storage (CCS) technologies are employed, a significant reduction in CO_2 per capita will be very difficult to achieve with coal-based systems, unless there is an increase in the use of clean fuels and a decrease in energy demands. Beyond the industrial sector, a number of actions are needed in Shanghai for low carbon energy transitions that are essentially 'urban' in nature, such as integrated urban spatial planning, a reduction in automobile dependency, the promotion of green buildings and co-generation systems, a dampening of the growth in overconsumption, and improvements in material cycle efficiency. These are often overshadowed by the presence of a large industry sector, but have strong path dependency. In effect, interventions that are seeking to move Shanghai, and other Chinese cities, towards a low carbon future face considerable challenges concerning the inertia of current energy systems and the limited potential for 'managing' transitions within existing infrastructure systems.

Conclusions

This chapter had four aims. The first was to show the urban contribution to China's energy uses and to highlight their importance in determining the energy and carbon profiles of the nation. The second was to show the internal dynamics of urban areas in relation to energy usage and CO_2 emissions at the meso scale by estimating and analysing these factors for the highly urbanized and economically most important cities. The third was to illustrate the historical transitions that have taken place in urban energy uses and CO_2 emissions and their drivers in urban areas at the micro scale through a detailed analysis of Beijing, Shanghai and Tianjin. The fourth was a more detailed analysis of emission scenarios for Shanghai and the issues involved in effectively managing a systemic transition. This multilevel analysis illustrates the considerable implications for key policies such as energy efficiency and climate change mitigation in China.

First, the analysis showed that the urban contribution to China's total and commercial energy uses can be estimated as 75 per cent and 84 per cent respectively in 2006. Since the energy use per capita in highly urbanized cities is rising and the rate of urbanization itself is rapid, it is inevitable that the urban contribution will increasingly determine China's energy uses and CO_2 emissions for the next few decades. Therefore, an effective response requires a comprehensive national strategy and guidance by the national government of cities regarding integrated planning for urban development and energy efficiency. This is especially important because urban development led by individual cities in China has been rapid, with few considerations for energy security and climate change mitigation.

Second, the analysis of the thirty-five cities in China that represent provincial capitals and cities mentioned in the national plan shows that they have a disproportionate influence on China's energy and economic activities. These highly urbanized and economically important cities had only 18 per cent of China's population but produced 40 per cent of CO_2 emissions in 2006. This shows that the population share of large cities may be small but that large cities' energy and CO_2 impacts are disproportionately large in the context of developing countries. This also counters the argument that larger cities are receiving unfair attention despite having a smaller share of the population. Instead, large cities should be a primary target for improving the energy security and for climate change mitigation in China. The energy-intensive cities are largely located in the central and western parts of China, which house energy-intensive industries and lie in climatically cooler areas. These urban regions deserve considerable attention as regards better technologies, more investment and improved urban energy systems and infrastructure.

Third, the analysis of the megacities of Beijing, Shanghai and Tianjin showed that 'high carbon' energy transitions have been under way in these major cities: the energy uses and CO_2 emissions have increased several-fold, with the industrial sector contributing the most in the past two decades, alongside rapid economic growth. It also revealed that the average urban energy use per capita in China is small, but in the case of key cities such as Beijing (11.9 tons/registered person), Shanghai (16.7 tons/registered person) and Tianjin (12.4 tons/registered person), these emissions are well above those for key cities in the developed world, such as Tokyo (5.9 tons/person for 2003; TMG 2006), Greater London (6.95 tons/person for 2003; TMG 2006) and New York City (7.1 tons/person for 2005, PLANYC 2007). The analyses showed that economic growth and energy intensity have played key roles in the increasing and dampening of CO_2 emissions, respectively. The role played by the fuel shift is nominal, but the demographic effect – the rapid increase in urban population – has played an important role in increasing emissions in cities such as Shanghai and Beijing, where there are large unregistered or floating populations. Cities' CO_2 emissions per capita have increased over the years but the carbon intensity of their economic activities has dramatically decreased in recent decades. Unfortunately, in the past few years the improvements in CO_2 intensity have either significantly slowed down or worsened, owing to the slow expansion of the tertiary sector and over-reliance on the traditional coal-based economy.

Finally, the focused analysis of Shanghai suggests that under BAU scenarios, energy demand in 2020 will increase by 3.6 times as compared with that for 2006. The CO_2 emissions from energy consumption in 2020 could increase to 640 million tons. This increase will create an enormous burden for energy security, environmental protection, CO_2 mitigation and energy conservation. If the current priorities and plans are successfully implemented, they could play a major role in reducing energy demand, improving air quality and mitigating CO_2 emissions in Shanghai. But if these plans are not implemented fully, then the energy and CO_2 reductions will be lower. Given that new climate targets are being discussed globally and many cities have already pledged large CO_2 reductions, Shanghai will come under pressure in the future to reduce its CO_2 emissions rather than slow their growth.

The economic imperative is overriding in cities in China, and in this period of rapid economic growth, Chinese city policies (many of which are essentially formed at the central level but implemented locally) have ample opportunities to improve the energy efficiency of industrial technologies, processes and facilities – the hardware side – and bring about the restructuring of cities' economies. Consequently, improving the energy intensity of city economies has become a major policy target since it is seen as mutually supporting other policy priorities, such as climate change mitigation, energy security and pollution control. However, such a policy, in effect, does allow energy use to increase and carbon emissions to rise for the foreseeable future.

However, the creating and managing of transitions to low carbon cities require a dramatic slowing down or an absolute reduction in the total volume of CO_2 emitted. The low-hanging fruit that cities in China are trying to capture at the moment, such as improvements in energy efficiency, represents an important first step in dampening the growth of emissions but is perhaps not enough in the immediate future for low carbon transitions. Beyond the industrial sector, a number of actions are needed in cities to bring about low carbon energy transitions. These are being implemented in different ways but are fragmented and often receive little attention.

First, strong policies to develop a better 'urban form' through devising an urban development framework are necessary. Such policies impose no serious costs but set development principles and principles for the spatial organization of cities, optimizing buildings, technologies, transport and infrastructure for energy. Cities are paying less attention to these overarching soft planning aspects.

Second, the role of cleaner fuel is important. The growth in energy use per capita is unlikely to decrease in the near future, but the use of more renewable energy and cleaner fuels would help to reduce CO_2 growth and develop societal and market systems based around cleaner fuels and renewable energy. City authorities can provide facilitation and support for the supply and use of renewable energy through financial incentives, piloting, demonstration and information provision to power companies, businesses and citizens.

Third, developing an efficient urban public transport system and discouraging the ownership and usage of private cars is critical. This is partly linked to integrated spatial planning but goes well beyond that. As a cause for concern, China sees the automobile sector as a strong pillar of the economy, and the prevailing urban sprawl implicitly helps to increase the demand for private automobiles. Shanghai has for many years had a policy of controlling vehicle numbers through a licence bidding system. Beijing has massively expanded its urban subway and has implemented mild control over the available parking spaces and the price of parking.

Last, but not least, the overconsumption of energy, goods and services which is spurred by income growth, subsidies and other factors needs to be dampened. City authorities can influence these factors through overarching policies and actions aimed at influencing lifestyle and behavioural change through correct price signals, information provision and educational campaigns. All these factors need to be taken into consideration in the planning of low energy and carbon transitions for Chinese cities.

Acknowledgements

The author would like to express appreciation for the editing, suggestions and comments by Simon Marvin and other editors of this book, which greatly improved the quality of this chapter. The chapter is based on two previous works by the author, published as Dhakal (2009) and Li *et al.* (2009), and on the author's contribution regarding city energy and carbon modelling reported (as a lead for

China analyses) as a chapter in *World Energy Outlook 2008* by the International Energy Agency.

Notes

1 See Dhakal (2009) for additional details of this research and research methodology.
2 These cities are Beijing, Tianjin, Shijiazhuang, Taiyuan, Hohhot, Shenyang, Dalian, Changchun, Harbin, Shanghai, Nanjing, Hangzhou, Ningbo, Hefei, Fuzhou, Xiamen, Nanchang, Jinan, Qingdao, Zhengzhou, Wuhan, Changsha, Guangzhou, Shenzhen, Nanning, Haikou, Chongqing, Chengdu, Guiyang, Kunming, Xi'an, Lanzhou, Xining, Yinchuan and Urumqi.
3 Note: while most energy is used in the transport sector, this sector is defined as Transport, Storage, Postal–Telecommunications Services, so some anomalies may occur regarding transport sector analysis.
4 Shanghai exercises control over vehicle supply in the city through Singapore-style bidding for a limited number of licences to own vehicles.
5 According to Dhakal (2009), Shanghai's total CO_2 emissions from energy use were 228 million tons, which included 27 million tons of CO_2 emitted upstream as a result of the use of imported electricity, 8 million tons of CO_2 emitted as a result of kerosene use in the transport sector (aviation) and 20 million tons of CO_2 from fuel oil use, some of which was for marine transport.

References

CESY (2007) *China Energy Statistical Yearbook 2006*, Beijing: China Statistical Press.

CSY (2007) *China Statistical Yearbook 2007*, Beijing: China Statistical Press.

Dhakal, S. (2009) 'Urban energy use and carbon emissions from cities in China and policy implications', *Energy Policy* 37: 4208–4219.

The Economist (2004) *The Economist*, 21 August: 55–56.

Global Carbon Project (2008) 'The Global Carbon Budget 2007'. Online, available at: www.globalcarbonproject.org/carbonbudget/index.htm (accessed 12 November 2008).

IEA (2008) *World Energy Outlook 2008*, Paris: International Energy Agency.

Li, L., Chen, C., Xie, S., Huang, C., Cheng, Z., Wang, H., Huang, H., Lu, J. and Dhakal, S. (2009) 'Energy demand and carbon emissions under different development scenarios for Shanghai, China', *Energy Policy* 38 (9): 4797–4807.

PLANYC (2007) *PLANYC: A Greener and Greater New York*, New York: City of New York.

Raufer, R. K. (2007) 'Sustainable urban energy systems in China', *New York University Environmental Law Journal* 15 (1). Online, available at: www1.law.nyu.edu/journals/envtllaw/issues/vol15/index.html (accessed 15 November 2008).

TMG (2006) *The Environment of Tokyo*, Tokyo Metropolitan Government Environmental White Paper 2006, Tokyo: Tokyo Metropolitan Government.

UN (2007) 'World urbanization prospects: The 2007 revision', United Nations Population Division of the Department of Economic and Social Affairs, New York. Online, available at: http://esa.un.org/unup (accessed 10 January 2010).

Wen, G. (2005) 'Cautions on China's urbanization', Commentary 11, January, Maureen and Mike Mansfield Foundation, Washington, DC. Online, available at: www.mansfieldfdn.org/pubs/pub_pdfs/wen0105_chinaurban.pdf (accessed 11 November 2008).

7 The 'eco-cities' Freiburg and Graz

The social dynamics of pioneering urban energy and climate governance

Philipp Späth and Harald Rohracher

Introduction

Proponents of transitioning towards more sustainable energy systems emphasize the significance of cities as important arenas for change. At this level, energy demand is organized, specific mitigation opportunities manifest themselves and communication with private energy consumers is possible. While megacities clearly have a higher impact on resource-related problems and climate change, small and medium-sized cities should not be neglected so far as their potential role as test-beds for ambitious new governance is concerned.

The cities of Freiburg, in Germany, and Graz, in Austria, have been perceived as ecological model cities since the mid-1980s, as is illustrated by various international awards (e.g. the European Sustainable City award for Graz in 1996) they have won, press coverage and the many visitors from other municipal administrations. Both cities are known for their ambitious environmental objectives and have played an important role in pioneering the search for concrete measures that can bring about transitions, particularly towards more sustainable and low carbon energy and transport systems. Social dynamics that have led to increased activism from local experts and city officials have emerged in both cities, resulting in ambitious and often successful pilot projects as well as national and international attention. However, this profile has never been without controversy. Particularly in Graz, much momentum seems to have been lost around the turn of the century. Furthermore, the actual achievements of both cities regarding emission reductions and energy efficiency never actually lived up to their forerunner image.

The cities of Graz and Freiburg can therefore serve as interesting examples of the dynamics and limitations of urban transition processes towards greater sustainability. In our contribution, we analyse the discourses and social dynamics of municipal climate policies and energy transitions in these two pioneering 'eco-cities' and explore which actors were able to shape and direct transition processes and to what extent. We are particularly interested in 'guiding visions' as a crucial element of transition processes, and in the relation of these initiatives and visions at a municipal level to the discourses and activities at regional, national and international levels. In concluding our analysis, however, we will explore whether, and to what degree, the initiatives in these two pioneering cities can be characterized as socio-technical transitions at all.

We find the cases of Graz and Freiburg particularly interesting as they seem to share some features that could make new forms of governance, for example concerning transition management, easier to implement: They are medium-sized, clearly arranged cities, yet host numerous competent experts, many of whom know each other well and meet frequently. They are also both characterized by a relatively active citizenry.

Differences in the legal frameworks and institutional set-ups of municipal policies in Germany and Austria are certainly smaller than, say, those between Germany and the United Kingdom (Bulkeley and Kern 2004). Even so, two main differences framing urban environmental politics stand out. First, the development of an energy policy action plan in Graz was required by provincial legislation (because of its air quality problems), whereas all of Freiburg's energy-related policies have been pursued voluntarily. Second, while the mayor of Freiburg is elected directly, the mayor of Graz is appointed by majority vote of the city council. Nevertheless, these differences in institutional frameworks have not resulted in any striking differences in policies or discourses.

In this chapter, we will particularly emphasize the example of Freiburg. We will use the case of Graz (for more details, see Rohracher and Späth 2009) more as a contrast to better work out the salient features of the change dynamics in the Freiburg case. In order to analyse the social dynamics of municipal climate policies and energy transitions in these two pioneering 'eco-cities', we ask: (1) What infrastructural, institutional or discursive structures have been changed? (2) Which actors were able to shape and direct such changes, and to what extent? (3) What role did guiding visions play, as guiding visions, according to the transition management approach, are crucial for intentional transitions? (4) What is the relation between municipal-level initiatives and discourses and activities at the regional, national and international levels? (5) Consequentially, to what extent can the initiatives in these two pioneering cities be called socio-technical transitions at all?

We draw our insights concerning the case studies from three types of sources. First, we draw on our own experiences as commentators and participants in the political debates about energy in Graz (H.R. from 1988) and Freiburg (P.S. from 1997). Additionally, we conducted a series of interviews in 2009 with major players and critics of these policies and discourses.[1] In order to triangulate our perceptions and findings, we conducted a brief analysis of press coverage on this topic (local and national newspapers and the municipalities' bulletins).[2] In the second section, we discuss the case of Freiburg and contrast it with findings from Graz, which are presented in the third section. In the fourth section of the chapter, we derive some tentative generalizations with regard to the role of urban discourses and social dynamics in socio-technical transitions resulting from the interaction of niches, regimes and a broader socio-technical landscape. We conclude with a summary and a note of caution concerning the ambition to manage processes similar to those analysed.

The case of Freiburg

The city of Freiburg has approximately 220,000 inhabitants and is located in the south-western corner of Germany. It is surrounded by the slopes of the Black Forest to the east and the upper Rhine valley to the west. It serves the region as an administrative and commercial centre and is characterized by an old university, limited industry and a highly service-oriented economy. Since the early 1990s, Freiburg has been praised by its inhabitants and visitors not only for its outstanding living quality but also for a set of remarkable environmental policies. The city has also received several awards (for further details, see Hagemann 2007: background section; Filtnib 2007; Hopwood 2007). Freiburg brands itself today 'the Green City', on the basis of a long history of pioneering solar energy and environmental policies. An important step in the formation of this profile was that Freiburg was awarded the title of 'Environmental Capital of Germany' in 1992 (Box 7.1).

Pioneering environmental policies

Widely acknowledged achievements in environmental policies in Freiburg can be traced back at least to the 1970s. As a result of an urban transport policy with a strong emphasis on environmental and social aspects, parts of the inner city were converted into a pedestrian zone in 1973. Since 1984, a monthly ticket ('environment protection card') has been offered, which put an end to decreasing passenger numbers. From 1982 to 1999, the share of public transport in the city's volume of traffic increased from 11 to 18 per cent and the contribution of cycling from 15 to 28 per cent; the distances travelled by motor vehicles decreased from 38 to 30 per cent. As early as 1986, the city council of Freiburg unanimously decided on an 'energy supply concept', prioritizing energy-saving measures (insulating buildings), strong efforts to increase the share of renewable energy (solar, wind, landfill gas) and a rejection of nuclear energy. About 50 per cent of

Box 7.1 **Freiburg, 'Environmental Capital of Germany'**

In 1992, Freiburg was chosen as Germany's 'Environmental Capital' for its pioneering achievements, such as the installation of an early-warning system for smog and ozone pollution, pesticide bans, recycling measures, for its transport policy and perhaps even, for its engaging 'green' image. Ever since, new innovations in the field of environmental protection and solar engineering have been achieved, which have, in turn, been accompanied by a series of awards: the European Public Transport Award, the German Solar Award, Federal Awards for Sustainability in Urban Development, the 'Sustainable Community' award of the Deutsche Umwelthilfe e.V. (German Environmental Aid Association).

Source: The City of Freiburg (2008): 'Freiburg Green City', available at www.freiburg.de/greencity

electricity is now generated in efficient combined heat and power (CHP) plants; there is one major plant at an industrial site and 90 small-scale CHP plants all over the city. The district heating system (partly fuelled with wood) has been systematically expanded to newly developed districts and now serves about 12 per cent of the city's population. Five wind turbines with a capacity of 1.8 MW each were installed within the boundaries of the city with the support of the city administration. Several small-scale turbines generate hydropower, and the capacity of photovoltaic generators now exceeds 10 MW.

A strategy to reduce greenhouse gas emissions was developed in 1996 with the goal of reducing CO_2 emissions by 25 per cent by 2010. The regular climate protection reports, which have been produced every two years since 2003, indicate that the targets are far from being met. Nevertheless, in 2007 the city council decided to commit the city to a CO_2 reduction of 40 per cent (compared to 1992) by 2030. After several years of debate, and following an initiative by experts and some city council members, the administration adopted twelve priority measures in 2007, and the city council decided that '10 per cent (€1.2 million) of the concessions that the regional power supply corporation Badenova AG pays to the city will be invested in climate protection projects, especially in the transport and building sectors', as the administration proudly reports in its 2008 brochure *Green City* (see Box 7.2).

The city has been supporting low-energy buildings since 1992, using a subsidy programme since 2002 and pioneering the enforcement of particularly high insulation standards: for any plot of land sold by the city, and for any development covered by a municipal development scheme, the city ensures by contractual arrangement that specific building standards exceeding national requirements by up to 30 per cent are met. These so-called 'Freiburg low-energy building standards', which the city council approved in 1992 and 2005, induce obligatory investment costs to exceed national requirements by about 5–15 per cent to begin with; the national standards regularly catch up after four to five years. In 2008, the city council decided to tighten the standards again: since then, passive house insulation standards have been required for all new buildings, with a few exemptions. The most environmentally friendly energy-supply option (e.g. district heating) is usually mandated contractually on the basis of an energy concept provided it can be realized with a cost increase of less than 10 per cent as compared to a standard option, usually individual gas heating. Perceiving itself as a forerunner in environmental policies, the municipality actively promotes this image, for example at world exhibitions (Expo 2000 and 2010) and through continuous PR efforts. The city has also participated in many competitions. In the most recent of these, the European 'Green Capital Award', it was selected as one of eight finalists, but ended up third after Stockholm (2010) and Hamburg (2011).

Not so easy: changing a city's hardware

While the city praises its achievements, the actual reduction of greenhouse gas emissions in Freiburg is not overwhelming. The decrease in CO_2 emissions by 13.8 per cent per inhabitant between 1992 and 2007 is mainly the result of a few

Box 7.2 The municipality of Freiburg on its climate policy and status as role model

Freiburg took climate protection seriously long before the issue was on the general political and economical agenda and as a result, in matters related to climate protection, Freiburg is nowadays considered a role model far beyond Europe.

Climate Protection Concept 1997

In 1996, the municipal council decided to reduce CO_2 emissions by 25 percent before the year of 2010. The successes achieved were remarkable. By applying a well-designed package of actions, emissions could be markedly reduced, particularly in the traffic and energy sectors. The share of nuclear power electricity was reduced by over half, from 60 to less than 30 percent. Almost 50 percent of the city's electricity is generated by combined heating and power plants.

Continuing into the future

Despite best efforts, Freiburg will presumably fail to reach its original goal of 25 percent less greenhouse gases by 2010. However, this is regarded as an incentive by the city to redouble climate protection efforts. In the summer of 2007, resulting from a climate protection report by the Eco-Institute in Freiburg, the municipal council decided to proceed with its climate protection concept and raised the benchmark even higher for the next phase: 40 percent less CO_2 by 2030. This goal is ambitious, but not unrealistic, since the national and international conditions for climate protection improved considerably last year.

Targeted investments

A strong local climate protection policy must by no means be limited to mere gestures or lofty declarations of intent. Realistic political and financial commitments are of huge importance. In the coming years, ten percent (1.2 million Euro) of the concessions that the regional power supply corporation Badenova AG pays to the city will be invested in climate protection projects, especially in the transport and building sectors.

Source: The City of Freiburg (2008): 'Freiburg Green City', available at www.freiburg.de/greencity

private investments. The biggest decrease came from a big CHP plant at a chemical production site, which replaced less efficient equipment and now feeds electricity and heat into the public grid. While several CHP plants had already been installed in municipal swimming pools and schools during the early 1990s, this project in 1998 and a further one in 2001 were the most recent extensions of CHP capacity to be realized.

While the city loves to call itself a 'solar city' in a 'solar region' and proudly presents the once ground-breaking solar power installations to visitors from abroad, the total capacity of 11.3 MW_{peak} in photovoltaic power plants (equalling 1 per cent of local electricity consumption) has long since been topped by other German cities of a similar size.

The district heating grid and tramlines have been extended to both of the major urban developments from the 1990s in the early stages of construction: to Rieselfeld (now home to some 9,000 inhabitants) and to the so-called model district Vauban (now with around 5,000 inhabitants). High-energy standards were prescribed for all buildings as part of the private contracts between the city as the final seller of the land and the developers. An obligation to connect all buildings to the district heating network was only disregarded in cases of passive houses with a particularly low heat demand and under the condition that the owners would invest in solar thermal plants (Späth 2005). All of the major decisions concerning these large developments were taken in favour of what was perceived as the greenest of all economically feasible options: extending the district heating network and tramlines, setting high insulation standards for buildings and so on. Although these decisions were fervently debated among experts and citizens, they were based only to a limited extent on serious and comprehensive assessments of the long-term impacts of these interrelated infrastructural decisions (ibid.). After these activities in the 1990s, not much followed except for the long-overdue refurbishment of some schools. These, however, were mainly induced by the increased availability of funding from national sources. There are also no major infrastructural projects in the pipeline that could be expected to significantly reduce greenhouse gas emissions in the near future.

An informal network of highly engaged actors

Freiburg is home to an extensive set of research organizations, companies, lobby organizations and individuals that deal with renewable energies (mainly solar energy) and environmental technologies and policies. Besides the Fraunhofer Institute on Solar Energy Systems (FHG-ISE) and the Institute for Applied Ecology, the International Solar Energy Society (ISES) has settled its world secretariat in Freiburg, along with the European secretariat of the International Council on Local Environmental Initiatives (ICLEI). Although Freiburg is traditionally not a site of industrial production, PV modules and, more recently, concentrated solar power plants not only have been developed but also are produced in Freiburg.

The density of individuals and organizations related to energy alternatives in Freiburg is concordantly explained with the history of successful resistance to having a nuclear power plant. In Wyhl, around 30 kilometres west of Freiburg, nuclear industry firms and the state government planned to build a nuclear power station in 1975. The construction site was occupied by a unique mix of local farmers, students from Freiburg and conservative citizens from the region and was developed into a camp of resistance, vision and self-instruction concerned with, among other issues, alternative energy. From this experience and milieu, several research institutes, lobby organizations and, later, companies were founded in the subsequent decades. For example, a regional association for renewable energy was founded in 1993 and, together with partners, set up a regional energy agency in 1999. Many highly informed activists who develop ideas for and comment on the city's environmental, energy and climate policies have been recruited from these energy-related institutions. Even the municipal administration that is officially in charge of these policies admits that most initiatives and exemplary pilot projects in Freiburg have been invented and accomplished by civil society actors, often amid significant opposition from officials.

These individuals are not formally organized, but are well informed on each other's activities, as they frequently meet on various occasions. Their vision of an alternative energy system and their positions on energy-related matters are shared by many citizens who are not professionally involved in energy issues but who are nevertheless well informed, as is proved by the floods of letters to the editor in the regional newspaper, or the success of some ad hoc collections of signatures. When experts, for example, voiced concern about the green electricity product of the regional energy provider Badenova, more than 4,000 customers of the company changed to a more ambitious and trustworthy provider with no relation to nuclear energy and the image of being an 'electricity rebel' (Energiewerke Schönau). This movement was absolutely uncoordinated – according to all of the usual suspects – and created enough pressure to make Badenova completely overhaul its green electricity product and switch to a more ambitious strategy of electricity procurement.

In general, Badenova plays an important role in implementing the political energy objectives of the city council, administration and citizens. The regional energy supplier was established in 2001 by merging the local suppliers of Freiburg and surrounding areas, securing a slight majority for the municipal shareholders while selling 47 per cent to the Bavarian gas supplier Thyga (owned by E-ON). With some of the money gained by this privatization, an Innovation Fund for Water and Climate Protection was set up: it is now endowed with €1.84 million per year from the company's profits. At the time, many local energy experts warned that this fund would have limited effects, but could distract from the fact that the company's policy would be dominated by shareholder value orientation instead of climate protection. However, it turned out that the new regional player, Badenova, appreciated the importance of fulfilling the regional customers' expectations of being a green and innovative energy supplier. Consequently, Badenova has engaged in many renewable energy and efficiency projects.[3]

Local discourse and political culture

The local discourse on energy issues is based on the widely shared view that nuclear power and fossil fuels are undesirable and that regional policies should aim to reduce dependency on these energy supply options. Even the most conservative or most liberal politicians and citizens agree that Freiburg should act as a forerunner in terms of energy and climate policy. Conflict regularly arises when car-users feel discriminated against (e.g. by the strict management of parking spaces), but the need for innovative and exemplary regulation with regard to energy-efficient buildings and energy supply goes largely uncontested. Furthermore, the citizens perceive themselves to a large extent as being in a position to actively shape the local parts of the energy system (as customers or owners of distributed generation capacity), and consequently also demand innovative action by their authorities. This may have to do in part with the collective experience of the socially heterogeneous and finally successful resistance to the nuclear power plant in Wyhl, as all of the interviewees in Freiburg assured us – and even, some imagine, with the historical tradition of resisting the authorities in the state capital, Stuttgart. As a result, this widespread attitude has led to an extraordinarily vibrant discourse among well-informed citizens and experts in the regional press and at regularly held events, which peaked again after the Chernobyl accident in 1986. Although this constellation is very challenging for the staff of the municipality, who do not always welcome and fully exploit this level of participation, the city obviously greatly benefited from it and proudly presents the outcomes of such debates to visitors and at international competitions and fairs.

The positive feedback loop of the green image

Since the mid-1990s, increasing numbers of visitors from Japan and a multitude of other countries have come to Freiburg, mainly to marvel at the environmental initiatives and successes. As a consequence, private actors and the municipality began to actively exploit this image of being Europe's showcase for environmental policy, and especially solar energy applications. Being awarded several prizes for innovative policies and continued efforts supported Freiburg in developing a green image. Probably more importantly, the marvellous landscape in sunny south-west Germany makes it very easy to maintain such a positive image despite actual achievements having lagged behind those of other cities for many years. As Bernd Dallmann, chief executive of the Freiburg Agency on Economy, Tourism and Trade Fairs (FWTM), puts it:

> Apart from science and technology, factors such as culture, climate, landscape, as well as the excellent quality of life in Freiburg complete the profile of the 'Green City' and consequently attract creative minds, investors and tourists from all over the world.

Many activists and observers of the scene in Freiburg agree that there seems to be a positive feedback loop: the image of the place, 'innovative and green', might attract particularly engaged individuals who, given the chance, opt to live and work in Freiburg; it may also strengthen organizations and networks actively working for exemplary projects and policies in the city.

Although the specific 'Green City' label was initially opposed by many individuals, local environmentalists now embrace the fact that the city officials present the city as a forerunner for sustainability. They know that they will have good arguments for their innovative projects and policy initiatives if the discrepancy between the shiny PR product 'Green City' and the actual achievements and policies becomes too great. The authorities would then have to face discontent from the demanding local population.

The case of Graz: similarities and differences

The Austrian city of Graz is only slightly larger in population than Freiburg and is located in a similar landscape on the southern slopes of the Alps. With around 255,000 inhabitants, it is the second largest city in Austria and both the capital and the administrative centre of the province (*Bundesland*) of Styria. However, the attractive location of Graz in a basin amid hills is a major disadvantage in terms of air quality and particulate matter levels. Culminating in the so-called smog winter of 1988/1989, this situation was one of the main drivers for the city to come forward with ambitious environmental targets and programmes, described in more detail below. Like Freiburg, Graz tried to position itself as an ecological model city and has been awarded various European and international awards for its activities and achievements in this field, such as the Greenpeace Climate Protection Award in 1993. Graz was the first European municipality to be awarded the International Sustainable City Award by the European Union in 1996 and – among others – also received the Sustainable Energy Europe Award in 2008.

Graz on the way to becoming an eco-city

As in the Freiburg case, initial environmental activities in Graz were most pronounced in the field of transport policy and started in the late 1980s and early 1990s. Graz was the first city in Europe to implement a speed limit of 30 km/h as a general regulation for the entire city area except for major roads. The centre has many pedestrian precincts, and the city administration put a lot of effort into promoting cycle traffic. Graz was also the first Austrian city to open a mobility centre and to change the public bus fleet to one using the more environmental friendly bio-diesel fuel. Only recently, Graz was awarded the title of 'Civitas City of the Year 2008' for its broad-portfolio sustainable transport policy.

Major policy changes took place after the 'smog winter' of 1988/1989, mentioned earlier. Since 1991, Graz has developed the first Austrian 'Local Agenda 21' Programme, the environmental programme 'Ökostadt 2000' (Eco-City 2000; approved by the city council in 1995) and, not long after, the 'Municipal Energy Concept', approved in 1995. Motivated by the municipal energy concept, Graz also became the first Austrian member of the 'Climate Alliance of European Cities' (Klimabündnis), with a greenhouse gas emission reduction target of 50 per cent by 2010 from a 1987 baseline.

These programmes and their sub-programmes, such as Ecoprofit, Ecodrive and Thermoprofit, gained considerable international recognition for several then-innovative features. Both the environmental and energy programmes had a comprehensive and integrative perspective involving various departments of the city administration and treating energy and environmental issues as cross-sectional policy issues. A second characteristic was the participative approach in the development of these municipal programmes, with stakeholder fora (including NGOs, companies, researchers, municipal utilities and others) and citizen participation processes as important elements of the programme design. Moreover, the documents produced were less like traditional policy documents with policy targets and measures for the city administration, but rather were action plans targeting a broad range of actors and making detailed suggestions for the implementation of various activities.

One salient sub-programme that was gradually adopted by many other cities internationally was the programme Ecoprofit, a corporate environmental protection programme offering companies consultancy services on the economic evaluation of environmental measures, certifying companies implementing these measures, awarding the best performers in this respect and, finally, creating a network with continuous activities and events for Ecoprofit companies. Company participation in this programme was high, and the idea of win–win measures and environmental protection out-of-profit motives was enthusiastically embraced by politicians and the city administration. A national 'Cleaner Production Centre' was set up in Graz and facilitated worldwide cooperation projects on Ecoprofit programmes between European countries and China, Tunisia and others. Recently, these international activities resulted in a public scandal for the city of Graz, with charges of fraud and other criminal offences and, consequently, the dismissal of the head of the municipal environmental protection department. The Ecoprofit programme in Graz, however, was not directly affected by these developments and continues to operate. Besides the Cleaner Production Centre, the various environmental programmes resulted in other institutionalization processes such as the establishment of a municipal energy agency, new procurement and energy-controlling regulations for the municipal buildings and so on.

However, these dynamics of urban environmental change seem to have significantly cooled down over the past ten years, both in terms of fewer new initiatives getting off the ground and, with a considerable time lag, also in terms of decreased public attention. This coincided with a general decrease in public attention towards climate policies in Germany (Bulkeley and Kern 2004: 37) and, arguably, also in Austria. Additionally, public disputes and resistance in recent years have evolved around air quality and the high levels of particulate matter, which damage the perception of Graz as a 'green city'. A question thus remains about the extent to which the environmental programmes and activities briefly outlined above have resulted in any kind of fundamental transition process for the city's energy system.

Characteristics of the developments in Graz

Space allows us to provide only a rough outline of the dynamics and background of the transition activities in Graz. While the process in Freiburg was highly driven by an active network of civil society initiatives, the developments in Graz were much more expert and administration driven. The external pressure arising from deteriorating air quality and the ensuing obligation (deriving from a Styrian provincial law) to produce an environmental action plan was seized by a network of people mainly within the municipal administration (such as the head of the environmental protection department and the head of the energy unit), politics (a city councillor and few members of the city assembly) and research (the Technical University, the research centre IFZ, etc.). Throughout further developments, some environmental consultancies played a more important role (such as the Austrian Energy Agency or a consultancy set up in connection with the Ecoprofit programme), but were closely linked with the already existing network. The central drivers, however, were the aforementioned municipal department heads, who took on an entrepreneurial role and mediated between policy, research and important stakeholders such as the municipal utilities. Most of the programmes such as 'Eco-City 2000' and the municipal energy and climate programmes or Ecoprofit were developed in close cooperation with the environmental department, but carried out by expert organizations with a strong research background and with the freedom to design a participative and comprehensive process. Several newly founded change agencies such as the Energy Agency of Graz, the Cleaner Production Centre and the private consultancy STENUM developed their own momentum and activities, and were largely unaffected by changes in city politics.

However, despite these seeds for long-term change, the momentum and public image of Graz as an eco-city has decreased since the end of the last century. Two of these core actors left the municipal administration – one as a consequence of his embroilment in a scandal focused on Ecoprofit's international activities, the other because of his feeling of not being sufficiently supported in his activities by the responsible city councillor. At the same time, public attention also shifted towards the role of Graz as the 'European Cultural Capital 2003', while urban energy politics were increasingly caught up in disputes about the privatization and reorganization of the municipal utilities.

In retrospect, the main success factors of environmental change in Graz, namely the entrepreneurial spirit of a few key persons in the city administration and the self-enforcing image creation and discourse about Graz as a model eco-city, turned out to be an Achilles heel for further development: the entrepreneurs left and the image was replaced in search of something fresher and newer. However, as much as the image and discourse exaggerated developments in its heyday, the current picture is probably gloomier than the reality of the environmental change measures in Graz. Not only is the environmental department better staffed and much more powerful than twenty years ago, but so too are the environmental initiatives in civil society. As part of a more diverse institutional framework of environmental action, new organizations such as the energy agency are now conducting work formerly done in the municipal department. Nevertheless, much of the momentum and support generated at the discursive level has been lost. The extent to which the entrance of the Green Party into the new city government in 2008 can bring new dynamics into the environmental transformation process remains to be seen.

The role of local initiatives in socio-technical transitions

We now want to relate these observations of discursive and institutional dynamics in the two cities to the growing body of literature dealing with system innovations and the transition of socio-technical systems towards sustainability. The dynamics of such transformation processes can best be understood in a multilevel perspective of innovation (Rip and Kemp 1998; Geels 2005). This perspective integrates a micro level of protected niches, functioning as test-beds for the emergence of new socio-technical constellations; a meso level of socio-technical regimes (such as energy systems); and the broader context of the socio-technical landscape, encompassing cultural norms, values and persistent socio-technical structures. The pivotal point in this concept is the meso level of the socio-technical regime, referring to the temporal stability of socio-technical configurations and meaning a rule set or grammar that structures the process of socio-technical co-evolution.

As arenas for transition processes, cities are somehow located at an intermediate level between niches and regimes. They are part of the energy regime, but at the same time are places where at least partial transformations can take place. In this respect, the notion of regimes conveys an exaggerated sense of homogeneity. On one hand, the regime level consists of the whole energy system (and its global extension); a fundamental energy transition would therefore have to take place at this scale. On the other hand, there can be rather significant variations of this regime between nations as well as between regions or cities, leaving room for manoeuvring at these levels to transform at least parts of the regime. With regard to the developments in our case-study cities, we can thus adopt a regime or a niche perspective: if we want to (1) put our analytical focus on the urban energy system and the respective (cultural) systems of governance and energy generation and consumption, we can consider these locally specified interrelations to be a meso-level 'regime'. Individual projects and policy initiatives are then the 'experiments' at a 'niche level' and not only relate to the development of the specific regime at municipal or metropolitan levels, but also are influenced by, for example, national energy policies and a general change in values at the landscape level. If we want instead to (2) focus on the contribution of urban developments to broader changes, we can consider the discourses and achievements in cities as being 'niche-level experiments' for transitions of an energy regime at a national level.

Indeed, most of the eco-city activities encountered in our case studies work at and mediate the niche and regime levels. On the one hand, they carve out niches for sustainable energy applications (sustainable city districts, employment of decentralized renewable energy sources, enforcing superior insulation standards, etc.). On the other hand, much of the work of actors within the city consists of embedding and stabilizing these changes in a broader regime context by, for example, aligning actors and facilitating institutional changes at a province or national level or by disseminating organizational innovations such as the Ecoprofit programme in Graz through networks of cities.

In our further analysis of the transition dynamics in our two case-study cities, we focus on the following three salient (and partly interrelated) issues: (1) the role of guiding visions in mobilizing and integrating activities; (2) characteristic actor constellations and dynamics of network formation as factors for successful transition processes at a municipal level; and (3) the interrelations between different levels of governance.

The role of visions

The development of guiding visions is a central element in concepts of managed transitions dealing with governance strategies to direct socio-technical change towards desired outcomes. Most of the analysed examples of historical regime shifts (such as the transition from coal to oil as the dominant energy carrier) have in fact not been intentionally steered by creating consensual collective visions (Berkhout *et al.* 2003; Meadowcroft 2005: 487). Nevertheless, diverse types of visions are always present in social action and can be an important element for stabilizing or destabilizing incumbent socio-technical regimes (Berkhout 2006). The extent to which socio-technical transformation processes are influenced by consensual political visions of future system states and the degree to which such processes can, despite their multilevel, multi-actor character, be intentionally managed is still an open and controversial question (Berkhout *et al.* 2003; Healey *et al.* 2003; Shove and Walker 2007).

Regarding our case studies, we can see that the actors share a vision of working towards a green and clean energy system. Especially in the case of energy futures, there is broad consensus about certain cornerstones such as increased energy efficiency, the use of renewable energy, the need to move away from nuclear and fossil fuel energy generation, and the promotion of sustainable lifestyles and behaviour. This general vision is rarely articulated in detail in terms of a specific socio-technical future to be realized at a certain point in time, specific pathways to be followed or trade-offs to be encountered. Instead, there is a general consensus about the direction in which to move. There were a few occasions in which ideas and objectives had explicitly been discussed in larger collectives such as energy-related Agenda 21 working groups or roundtable discussions (EnergieTische). Informal communication seems to be sufficient to align the action of many actors, so they do not feel a lack of consensus with regard to requirements, but rather a lack of means to implement the consensual measures.

On the other hand, programmatic documents such as municipal energy and climate protection plans often work with rather detailed guiding visions about sustainability targets, pathways by which to get there, policy measures and so on.

In some cases, such as with the municipal energy and climate plan in Graz, this vision has even been collectively developed, involving city officials as well as stakeholders and the wider public. In terms of the ideas of transition management, these are exactly the guiding visions needed to orient and integrate collective action. In our case studies, however, this orienting and integrating role was instead played by the implicit, general visions mentioned above.

Actor constellations and networks

An interesting element for better understanding the eco-city activities or transitions in Freiburg and Graz is the specific constellations of actors and their dynamics. In both cases, transition activities were not policy-driven processes but largely dependent on new constellations and cooperation across different types of actors and stakeholder groups. The process in Freiburg was largely driven by a network of civil society organizations, including environmental research institutes and environmental intermediaries such as the Ökoinstitut (Institute for Applied Ecology) and ICLEI. Administration and politics gave way to this push and tried to make the best of it by taking up the initiatives and marketing an ecological image of Freiburg to the public.

The process in Graz was much more dependent on entrepreneurs within the administration and their links with research institutes and other expert organizations outside the city administration. In this case, a rather small network of actors across politics, administrations and research institutions was able to set the agenda for a transition process and create a dynamic that increasingly brought politicians and other actors (e.g. corporate actors) on board and could also enrol media and attract public attention locally, nationally and to some extent even internationally. The Graz case also points out the danger of such networks disintegrating again and the importance of institutionalization processes to keep the transformation process going. Although the public profile and level of activities have decreased in Graz over the past ten years, institutions such as the Energy Agency have helped existing programmes to continue and have even started new initiatives and improvements.

In both cases, transition management was more about creating pressure for politics, generating sufficient public attention and forging actor-networks strong enough to mobilize substantial resources for environmental activities. Without a doubt, more consistent and effective changes could have been achieved with a stronger role of politics and better planning and implementation strategies. However, the contingent and piecemeal character of the process observed in both cities is most likely more characteristic of transformation processes in reality. What can probably be generalized from both cases is that a variety and certain density of research, civil society organizations and strategic environmental intermediaries were important resources for exploiting emerging windows of opportunity for energy transitions, whether as a reaction to external pressures or the emergence of actor constellations striving to transform municipal energy policy.

The interrelation of municipal processes with other levels of governance

Although the discourses on environmental policies in Graz and Freiburg focus heavily on opportunities at the municipal level, the main actors do consider the repercussions of their local actions at other levels of governance. If the head of the regional energy agency in Freiburg sees a need to protest against a cutback in a national support scheme for energy efficient refurbishment, he quickly mobilizes a so-called climate alliance of the solar region of Freiburg, which, in this example, translates into a list of thirty-nine local organizations supporting his open letter to the federal minister for economic affairs.

The relatively small cities studied are probably not important enough to try to get special deals, or even a mandate to manage transitions, as has been observed in the case of world cities such as London (Hodson and Marvin 2009). Nevertheless, both cities are highly active in national working groups and representative bodies of municipalities, and also participate in international networks of municipalities that try to put pressure on decision makers at national and international level, such as ICLEI, the climate alliance and the covenant of mayors – a manifesto for intensified climate protection. We know of no attempts by the two municipalities to directly influence federal legislation, but they explicitly aim to act as national models, for example by enforcing superior building standards, and thus influence the national debate on what is achievable.

The cities see themselves as being in a national or international competition to attract economic investments, particularly in 'clean' companies, as well as organizations' headquarters and tourists. Administrations promote a green image as an asset in this competition. Some national and international organizations exploit these tendencies and try to organize a race to the top with regard to environmental standards and policies (see, for example, the European Green Capital Award). Sustainability policies of municipalities are also appreciated and strategically supported because of their educational function, playing on the local government's relatively direct contact with the mood of its citizens.

Regarding the contribution that smaller cities can make to socio-technical transitions, our cases indicate that the relationship to the state level is of particular importance. The success in reshaping the local energy system configurations in both case studies has been limited, because of a lack of cooperation with and support by the government at the state level. In Freiburg, for example, the local efforts to develop sites for wind turbines were sometimes blatantly opposed by authorities at the state level, and the buildings and heat generation capacities under state control were not accessible for a synergistic integration into a joint heat network. While a wish for cooperation seems to exist in the municipality of Freiburg, it does not yet exist in relevant parts of the government at the state level. In Graz, there also seems to be a feeling that the capital city should develop an energy policy independently of initiatives at the provincial level. We see that a culture of non-cooperation can be rooted in various preconditions. A very recent debate about the planned relaxation in the regulation of urban sprawl in the countryside around Graz reminds us that the municipality has little influence, even on matters that strongly affect the immediate neighbourhood of the city and what is achievable for the agglomeration in terms of sustainable land-use patterns and infrastructure.

We can conclude that the potential contribution of cities to broader transitions is highly dependent on supportive framework conditions. Actor networks probably need to span beyond the municipal level for substantial changes to be achieved, and the municipal dynamics can only be fully exploited if resources (money, authority) allocated to the municipalities from central governance levels are not too restricted. If these conditions exist, however, even smaller cities could play an important role as test-beds and showcases of new configurations, also having a significant impact on national and international discourses and policies.

Conclusions

In this chapter, we have analysed the dynamics underlying the largely successful positioning of the cities of Graz and Freiburg as environmental forerunners. These municipal-level processes turn out to have been driven by a heterogeneous set of non-governmental actors or by entrepreneurs within the city administration rather than by strategic actions from municipal governments. These actors aligned themselves on the basis of localized, informally shared, yet remarkably stable visions of a sustainable energy future in the cities. The vision implies that certain actions are to be taken locally and envisages that they contribute to the desired general shift of the energy regime at the global level.

Various extra-governmental policy entrepreneurs were able to exploit windows of opportunity for the mobilization of resources, such as shifts in public opinion in favour of environmental policies (the smog winter in Graz, an anti-nuclear struggle in the 1970s and Chernobyl for Freiburg). In many incidents, expert activists were supported by an interested and demanding citizenship expressing their preferences, for example in letters to the editor of the local press, perhaps influencing officials to finally adopt innovative policies and programmes.

In the case of Graz, municipal energy policy has been strongly shaped by an informal network of a group of individuals who had previously been committed to sustainable energy options at the Technical University. Eventually, some of these individuals assumed central positions within the municipality and, for a limited time, had a strong influence on municipal programmes and policies. However, as it turned out, such a narrow network, focused largely on individuals within the administration, is more vulnerable to disruptions such as the loss of a few central individuals. In the case of Freiburg, the network was more deeply rooted in civil society and more widely distributed across societal spheres. The main experts and activists could not, or did not want to, join the administration itself. Relationships between municipal actors and external experts therefore remained relatively distant. The different dynamics of the energy transition processes in Freiburg and Graz can thus be at least partly explained by the different morphologies of the underlying policy networks, as illustrated in Figure 7.1.

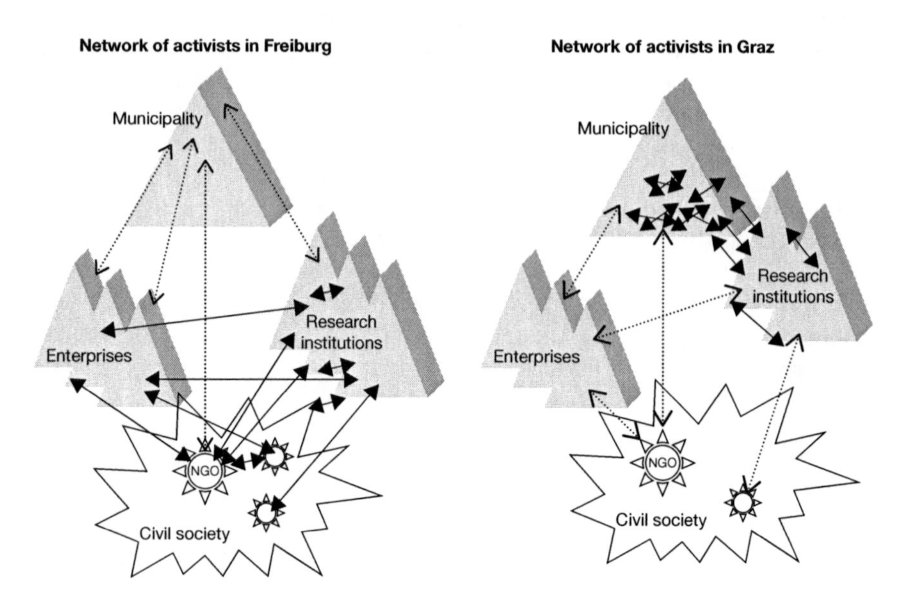

Figure 7.1 Stylized visualization of the social networks in Graz and Freiburg.

In both cases, there was clearly no overarching and integrative process for setting long-term targets, implementing activities and monitoring. If there had been such strategizing, many contingent factors, actor constellations and so on. would probably still have played quite an important role. In the light of these findings, the ambitions of the transition management approach appear to be exaggerated. Manageability, in the sense that sticking to long-term plans seemed possible and advantageous, remains an illusion, even in these rather small and privileged cities.

Nevertheless, the case studies leave room for the assumption that if some conditions are met, even cities of little economic and political power can gain great importance for systemic change in socio-technical constellations as places of innovation and experimentation (e.g. with regard to regulation) and as aggregations of energy customers with significant market power. At this governance level, the generation of distributed energy can be well supported, and where it has been institutionalized, municipal influence on regional energy companies can contribute significantly to influential experiments with alternative socio-technical constellations.

However, several crucial preconditions seem to exist. First, certain windows of opportunity for social mobilization, such as the Chernobyl experience or price shocks, need to be exploited and translated into an articulate demand for superior environmental policies. Such mobilization, second, will be pursued successfully only if actors from various segments of society and at different levels of governance feel that their individual interests overlap. Furthermore, municipal officials could probably make a significant difference if they were more appreciative of, and actively tapping into, the knowledge, relationships and other resources that non-governmental actors could bring into a joint pioneering venture in energy and climate policies.

Notes

1 In Freiburg, seven individuals were interviewed: a staff of the municipal energy department, the head of the regional energy agency, three energy experts critically involved in the municipal debates, a consultant on urban governance issues and a specialist journalist. In Graz, four interviews were conducted, with the current head of the municipal environmental department and former key persons in the energy policy process and NGOs.
2 Databases have been searched for certain terms (e.g. Freiburg AND Eco- City/solar region/green capital/environmental capital) and the relevant hits (for Freiburg, for example, 419 articles from thirty-seven newspapers and magazines) were compiled and selectively coded.
3 Since many customers and citizens do not appreciate that the standard electricity product of the company still contains about 20 per cent nuclear power and that nearly half of the profits made by the company might end up with E-ON (the big 'non-green' energy agglomerate), there are now discussions to use the pressure put on E-ON by the anti-trust authorities and bail out the shares held by Thyga/E-ON by means of an increased involvement of municipalities, a consortium of regional energy companies and even private shareholders.

References

Berkhout, F. (2006) 'Normative expectations in systems innovation', *Technology Analysis and Strategic Management* 18 (3/4): 299–311.

Berkhout, F., Smith, A. and Stirling, A. (2003) 'Socio-technological regimes and transition contexts', SPRU Electronic Working Paper Series 106. Brighton: SPRU (Science & Technology Policy Research).

Bulkeley, H. and Kern, K. (2004) *Local Climate Change Policy in the United Kingdom and Germany*, Berlin: WZB.

City of Freiburg (2008) 'Freiburg: Green City', City of Freiburg. Online, available at: www.freiburg.de/servlet/PB/show/1199617_l2/GreenCity.pdf (accessed 11 May 2010).

Filtnib (2007) Visitor's blog, praising 'the greenest city of them all'. Online, available at: http://filtnib.com/2007/04/10/europes-greenest-city/ (accessed 11 May 2010)

Geels, F. (2005) *Technological Transitions and Sysem Innovations: A Co-Evolutionary and Socio-technical Analysis*, Cheltenham, UK: Edward Elgar.

Hagemann (2007) The 'background' section of a case study on Freiburg. Online, available at: www.pvupscale.org/IMG/pdf/Schlierberg.pdf (accessed 11 May 2010).

Healey, P., de Magalhaes, C., Madanipour, A. and Pendlebury, J. (2003) 'Place, identity and local politics: Analysing initiatives in deliberative governance', in M. A. Hajer and H. E. Wagenaar (eds) *Deliberative Policy Analysis. Understanding Governance in the Network Society*, Cambridge: Cambridge University Press.

Hodson, M. and Marvin, S. (2009) 'Cities mediating technological transitions: Understanding visions, intermediation and consequences', *Technology Analysis & Strategic Management* 21 (4): 515–534.

Hopwood, D. (2007) 'Blueprint for sustainability? What lessons can we learn from Freiburg's inclusive approach to sustainable development?', *Refocus* 8 (3): 54–57.

Meadowcroft, J. (2005) 'Environmental political economy, technological transitions and the state', *New Political Economy* 10 (4): 479–498.

Rip, A. and Kemp, R. (1998) 'Technological change', in S. Rayner and E. L. Malone (eds) *Human Choice and Climate Change: Resources and Technology*, vol. 2, Columbus, OH: Battelle Press.

Rohracher, H. and Späth, P. (2009) 'The fragile dynamics of urban energy system transitions: The eco-cities of Graz and Freiburg in retrospect', Paper given at International Roundtable Conference 'Cities and Energy Transitions: Past, Present, Future', Autun, France, 1–4 June.

Shove, E. and Walker, G. (2007) 'Comment: CAUTION! Transitions ahead: politics, practice, and sustainable transition management', *Environment and Planning A* 39 (4): 763–770.

Späth, P. (2005) 'District heating and passive houses: Interfering strategies towards sustainable energy systems', Paper given at ECEEE 2005 Summer Study proceedings, Stockholm.

8 The rise of post-networked cities in Europe?

Recombining infrastructural, ecological and urban transformations in low carbon transitions

Olivier Coutard and Jonathan Rutherford

Introduction

More than twenty years ago now, in the foreword to their seminal collection of essays on 'the rise of the networked city in Europe and North America', Joel Tarr and Gabriel Dupuy pondered 'to what extent the cities of the future will continue to depend on the infrastructure technologies of the nineteenth century, and to what extent they will incorporate new and more flexible technologies' (Tarr and Dupuy 1988: xvi). This question has gained a particular salience in recent years. Indeed, we are witnessing an unprecedented critique of the extensive networked infrastructures built over the past 150 years for the provision of essential services such as water, sanitation, electricity and heating. In the wake of (and in response to) this critique, there is a parallel rise of alternative, smaller-scale technological systems, which are frequently viewed as more 'sustainable'. While convinced that large infrastructure networks will long into the future continue to play a crucial role in basic service provision to urban populations around the globe, we focus in this chapter on the development of socio-technical alternatives to these large systems and on the new combinations between the former and the latter that result from this development.

This dynamic is of particular importance for urban futures in a context of increasing concern for climate change mitigation (and adaptation) and energy security in the face of diminishing fossil fuel availability. The promotion of alternative, decentralized technologies is viewed by activists, experts and policy makers on all levels as a promising pathway to low carbon cities, even though it remains an open question as to whether and under what conditions urban planning increasingly based on these technologies would be more likely to promote low carbon urbanism than planning based on traditional, centralized infrastructure networks.

The development of 'small-scale', 'decentralized', 'dispersed', or otherwise 'alternative' technologies clearly problematizes the inherently networked nature of the urban, on the environmental, spatial, social and political levels that technical infrastructure always implies and impinges on. In this chapter, we explore the nature and widespread implications of this nascent infrastructural transition in energy provision and beyond. The chapter examines some of the contradictions and tensions present in these alternative systems in order to critically interrogate the shifting interplay between the 'technical' and the 'urban'. Through this focus, we seek to move beyond simplistic and deceptive ideas that these alternative, decentralized technological solutions are always inevitably more 'beneficial' and 'sustainable' than traditional networked urban infrastructures. In doing this, we propose the notion of a 'post-networked city' as a way of designating the myriad of emerging urban-infrastructural configurations in a post-'networked city' era.

We proceed in four stages. First, we discuss existing approaches to cities, urban infrastructures and socio-technical change and highlight the need to get beyond a dominant sustainability rhetoric upon which critiques of networked urbanism are often based. We set out an alternative framework for exploring the forms, implications and tensions inherent to the emergence of decentralized technological systems. In the second section, we propose a typology of 'post-networked' pathways – especially to proclaimed low carbon urban futures – and then, in the third section, begin to analyse the shifting patterns and wider urban implications and contradictions of their development. A concluding section considers the extent to which different, perhaps incompatible, conceptualizations of decentralized technological service provision – in relation to low carbon transitions and beyond – are being mobilized by different groups and identifies some of the issues that both researchers and public authorities will need to be closely attentive to.

Cities, technological systems and the sustainability imperative

Because large networked infrastructure systems have been shown to be so central in the development and functioning of 'modern' cities (Tarr and Dupuy 1988; Kaika and Swyngedouw 2000; Hughes 1983), it is reasonable to assume that any weakening or transformation of these systems will affect the urban condition. This is important because the critique of the 'networked city' paradigm in European societies is more widespread, more radical and more systematic today than at any time since the emergence of large, centralized urban infrastructure systems from the mid-nineteenth century. This critique has emerged from the combination of two main processes.

First, from the late 1970s a quadruple movement of political, economic, technological and socio-spatial decentralizations, combined with a 'decollectivization' of certain societal functions and meanings, has increasingly challenged the traditional forms of organization and governance of large networked systems. This evolution has included the emergence of more polycentric and multilevel forms of political governance; the application of neo-liberal reforms, including privatization and liberalization; the technological 'stasis' (Hirsh 2003) of centralized infrastructure and the concurrent downscaling of systems of service provision; and the reinforcement of processes of socio-spatial segregation and 'individualization' of the urban experience.

Second, these overlapping shifts have combined with the context created in particular over the past decade by the rising imperative of 'sustainable (urban) development', which rests on ideologies and imaginaries that promote both the (re)localization of urban metabolisms and a concern for urban resilience and autonomy in terms of reducing certain vulnerabilities and risks (e.g. security of resource provision) (Hodson and Marvin 2009; Newman *et al.* 2009).

We argue that what is at stake is a radical and systemic critique of the network paradigm, which is viewed as contradictory to the allegedly more sustainable techno-ecocycle paradigm. The network paradigm is based on a principle of linear flows: resources are abducted 'upstream', more or less remotely from the (urban) consumption centres where they are transformed, and all sorts of wastes resulting from this transformation are rejected in the environment 'downstream'. Fundamentally, the network paradigm rests on the historically constructed conception that it is always advantageous to expand the network: to satisfy what is postulated to be an ever-growing demand, to improve quality and reduce costs, and even to solve the problems associated with the expansion of the network. This strongly contrasts with several usually emphasized tenets of sustainability: circularity, autonomy and sobriety. Table 8.1 details the diverse levels of opposition between the 'network' and the 'techno-ecocycle' paradigms.

Based on this critique, various forms of 'decentralized technologies' (DTs) have flourished in the provision of environmental services (energy and water provision, sanitation, waste management, etc.). These DTs are often perceived – and indeed constructed – by experts, policy makers, service providers and other stakeholders

Table 8.1 Contrasted (opposed) paradigms: large technical networked system versus sustainable 'techno-ecocycle'

Large centralized network	*Sustainable techno-ecocycle*
Linear metabolism: tapping > supply > disposal	Circular metabolism: recycling, reuse, retrieval
Decoupling between local resource availability and use	(Re)coupling between local resource availability and use
Territorial solidarity	Territorial autonomy
Technical systems	Ecological systems
Flows, imperviousness, kinetics	Stocks, porosity, stasis
Hydraulics-based model	Resource-based model
Supply-side model	Demand-side model
Economics of expansion and growth (scale, scope, club)	Economics of preservation
Unbounded, ever-growing consumption	Bounded consumption
Sector-based, sequential management	Cross-sector, integrated management
Irreversibility, obduracy, 'momentum'	Reversibility, adaptability
Carbon dependent	Carbon neutral

Source: Adapted from Coutard (2010: 114).

on all levels as more coherent with the sustainable techno-ecocycle paradigm (see Greenpeace 2005; Patterson 2005; Willis 2006; Gilroy-Scott 2007). In a recent report aimed at promoting the development of decentralized energy in London, for example, the executive director of Greenpeace accused UK government policy (oriented towards centralized fossil fuel and nuclear generation) of being 'fixated with the technologies and infrastructure of the past' (Mayor of London and Greenpeace 2006: 3). As if to echo this critique, London's then deputy mayor, Nicky Gavron, emphasized in her foreword to the city's 2007 Climate Change Action Plan that 'we are spearheading a decentralised energy revolution here in London [because] remote centralised power stations are the primary cause of climate change' (Mayor of London 2007: vii). The issue, normatively speaking, is that the shift from centralized networks to decentralized systems can be interpreted in diametrically opposed ways, which significantly problematizes this equation.

On the one hand, recent critical perspectives on the evolution of urban infrastructure have focused on the diverse ways in which infrastructure provision has become bound into the restless workings of global capitalism, as the very embodiment of the urbanization of capital investment (see Harvey 1985), with its privileging of profit and performance over equality and solidarity. In analysing the relative decline of 'the modern infrastructural ideal' and the subsequent rise of new political economies underpinning networked urbanism, Graham and Marvin's (2001) pioneering work, for example, firmly decries the technico-organizational unbundling and fragmentation of networks, and the socially regressive repercussions of this in cities around the world.

On the other hand, an overlapping literature has developed on the continuities and discontinuities between the development, maintenance and possible decline of 'Large Technical Systems' (LTS) of infrastructure and the making, learning and managing of 'new' technical solutions demanded by the urgent need to shift societies on to more sustainable pathways of development. In this literature, we see a shift from a direct focus on big society-changing technologies (e.g. the urban-regional energy systems of Thomas Hughes) to study of how and why innovative 'niche' technologies break out (or not) into wider society. This recent work on socio-technical transitions has done much to analyse the systemic change dynamics associated with LTS and, in particular, with the adoption and diffusion of 'decentralized' technologies (see, for example, Elzen *et al.* 2004; Geels 2002; Rotmans *et al.* 2001; Smith *et al.* 2005), usefully (re)focusing attention upon the smaller-scale technical evolutions, which many of these authors view as more sustainable socio-technical alternatives to the traditional, centralized LTS of the modern era.

The normative orientations of these two approaches are largely opposed. The splintering urbanism approach denounces the dystopian implications of the 'unbundling' of large networked systems, reflecting an inherent bias of 'networked cities' researchers in favour of large centralized infrastructure. There is a normative assumption that it is this form of infrastructure provision which always corresponds most effectively to the promotion of urban cohesion and equality. Meanwhile, the transitions literature dominantly praises the potential for improved sustainability inherent in socio-technical transitions and the advent of alternative, decentralized systems. This emerging literature, which has only rarely been sensitive to questions of space and place, appears to have partly lost the socio-political sensitivity of earlier LTS work (Hughes 1983; Summerton 1994; Coutard 1999) in favour of a more or less 'pure' theorizing of innovation (see also Shove and Walker 2007).

Neither approach, therefore, is fully adequate for exploring the current and emerging forms of basic service provision which we are interested in here. In particular, we argue that either by avoiding any evaluation of fragmented systems of service provision (splintering urbanism) or by contributing to normative policy views of their inherent sustainability (socio-technical transitions), both approaches have prematurely closed or too easily bypassed debates on the 'effects' of the rise of decentralized technologies. Our aim is therefore to (re)open ways to critically analyse the variety of forms and ambivalent implications of the 'post-networked city'.

Our position derives from our understanding of systems of utility service provision as Large Technical Systems. These systems encompass four main dimensions. First, they are high-cost systems (especially as regards investment costs) in which the issue of financing has always been crucial (Hughes 1983). Second, they are intrinsically territorial systems, both territorialized and territorializing – that is, affecting the organization of territories (Dupuy 1985). Third, although under-explored in much of the LTS literature (Joerges 1999: 282 ff.), Large Technical Systems fundamentally shape the ecological metabolism of cities and societies, as is strikingly shown in many historical cases (see, for example, Tarr 1997; Melosi 2000). Fourth, these systems have always had close connections with institutions and government at city, urban-regional, national and even supranational levels (see Offner 2000; Lorrain 2003 on the urban level). In discussions of the actual or potential transformations of these systems, we therefore need to consider the financing, socio-spatial, ecological metabolism and governability implications of alternative, decentralized forms of service provision in order to highlight the contradictions and tensions that are inherent in their development and that affect their potential as an instrument for the promotion of low carbon transitions. The pertinence of these various means and outcomes of systemic (infrastructural) change is empirically illustrated in the next section, where we explore some of the forms of post-networked urbanism currently emerging in European cities, and is then discussed analytically in the section after that.

Pathways to low carbon futures? A typology of emerging forms of post-networked urbanism

Challenges to large networked systems occur in various forms. In this section, we develop a typology of these forms based loosely on (1) the mix of social, political/institutional and technical presuppositions and discourses mobilized to promote decentralized technologies; (2) the tools, instruments and mechanisms through which these technologies are or are planned to be rolled out; and (3) the ambivalent implications and outcomes of these processes and practices. Although the cases analysed constitute exemplary, emblematic examples of each type, we argue that they also represent broader tendencies towards post-networked urbanism in the diversity of forms illustrated here (Table 8.2).

Off-grid processes

Perhaps the most radical form is based on a deliberate policy or collective strategy of bypassing to some extent traditional centralized networks and developing services on a local level, increasingly over decentralized, local infrastructures. Such policies or strategies are founded on desires or obligations for autonomy or independence, and effectively create delinked 'islands' in the form of local communities 'left' more or less 'to their own devices' for basic service provision. Even though our focus here is on 'delinking' *policies*, it is important to note that 'off-grid' processes vary in scale (household, community, municipality, etc.) and do not always result from innovative public policies. Individual disconnections can also increasingly be observed, whether from financial, health or political motivations (see Montginoul 2006).

The Woking experiment

From the early 1990s onwards, the Borough Council of Woking, a commuter town 45 kilometres south-west of London with a population of 90,000, has been at the forefront of best practice in local energy policies by initiating and developing actions based on local production and distribution, thereby promoting autonomy in energy provision with regard to the national grid. From its initial focus on energy efficiency, the council moved on in the late 1990s, through the drive and encouragement of its senior officer and chief financial officer, to establish its own energy services company (ESCO), Thameswey Energy Limited, which would own and operate a plant for the production and supply of electricity, heat and chilled water to customers, and develop and implement technologies for the production and supply of energy (Woking promotional material: Thameswey Energy Ltd,

Table 8.2 A typology of post-networked pathways to low carbon transition

		Organization	
		Delinking, unlinking (local autarky)	*New forms of linking (local autonomy)*
Decision	Collective, local scale of decision	A. Off-grid	B. Loop closing
	Individualized (or quasi-individualized) scale of decision	C. Beyond net	D. Feed-in to grid

November 2007). Setting up Thameswey as a public–private joint venture allowed it to bypass central government capital controls that restrict the scale of local government investments and projects. Thameswey has therefore used mostly private finance to build and operate a number of local community energy projects, including a small-scale combined heat and power (CHP) heating and heat-fired absorption cooling system (see Figure 8.1), a private wire renewable energy system delivering supply direct to council-owned housing and town centre businesses, and the first commercially operating fuel cell CHP system (South West Renewable Energy Agency 2007: 1–2). This decentralized system operates autonomously, even though it remains connected to the national grid (as an emergency backup): the Holiday Inn hotel in Woking, for example, was actually built without grid supply. But national regulation in turn limits the size of the local system and the number of customers that can be supplied (South West Renewable Energy Agency 2007: 4). Overall, the Woking project has made considerable economic, energy and emissions savings, which resulted in Woking being awarded local government Beacon status three times in four years, for Sustainable Energy (2005–2006), Promoting Sustainable Communities through the Planning Process (2007–2008) and Tackling Climate Change (2008–2009).

From Woking to London

Allan Jones, the senior officer who set the ball rolling and supervised the policy, was headhunted by the then mayor of London, Ken Livingstone, to become CEO of the London Climate Change Agency when it was established in June 2005 with the task of 'doing a Woking for London'. The national exemplar in this case has been an ordinary city which the big world city has been seeking to replicate. The energy policy initiated by Ken Livingstone, and not so far greatly contested or overturned

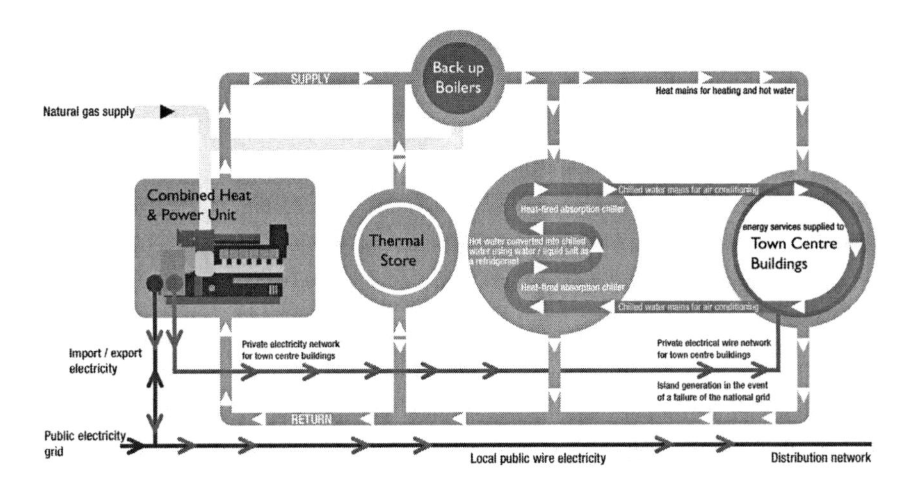

Figure 8.1 Woking Energy Station off-grid CHP system.

by the current mayor, Boris Johnson, rests upon 'a decentralised energy revolution' (Mayor of London 2007: vii) on both a metropolitan and a more local level. On a metropolitan level, the mayor and the Greater London Authority's strategy is nothing less than to delink London from UK national centralized energy supply:

> The Mayor's top priority for reducing carbon emissions is to move as much of London as possible away from reliance on the national grid and on to local, lower-carbon energy supply (decentralised energy, including combined cooling heat and power (CCHP), energy from waste, and onsite renewable energy – such as solar panels) . . . to enable a quarter of London's energy supply to be moved off the grid and on to local, decentralised systems by 2025, with the majority of London's energy being supplied in this way by 2050.
>
> (Mayor of London 2007: xxii)

On a local level, the London Development Agency is working on delivering decentralized energy through CHP plants linked to modern efficient community (district) heating networks, for example in east London. The borough of Southwark has integrated local water and energy provision ('sustainable community infrastructure') into its Elephant and Castle regeneration project, and has set up a multi-utility services company (MUSCo) for the delivery of this local infrastructure, for example 'private wire' energy, which 'will operate in parallel to the grid'.

Loop closing (circular metabolism)

This second form of challenge to large networked systems is based on a similar reasoning to the previous 'off-grid' form, but instead of promoting technical 'autarky' within the political boundaries of a community, town or city, the aims here are more directly ecological (or 'metabolic'): the objective is not only to locally produce water and energy and treat waste water and storm water, but also to reuse and recycle local waste and waste water in this production/treatment, and therefore, as much as possible, to move from a linear (production–consumption–waste) model to a model that closes the loop between these previously distinctive 'stages' in the life cycle of water and energy resources.

Hammarby Sjöstad

The emblematic Hammarby Sjöstad project in Stockholm is an urban regeneration initiative conceived in the early 1990s as part of Stockholm's failed bid for the 2004 Olympics which has transformed an old port and industrial area just to the south of the city centre into a contemporary residential and work area that will host some 13,000 apartments by 2015. In recent years, it has been a cornerstone both in the municipality's plan to build 20,000 new apartments and in the vision of creating an emblematic 'sustainable' district that would boost the visibility of Stockholm on the global stage. When work started, in the mid-1990s, planners were able to build on the fact that the City of Stockholm owned the water, energy and waste companies, which had been instructed by municipal politicians to work together with the planners on a neighbourhood-level recycling model for the whole project. One of the cornerstones of the project came to be, therefore, the creation of a recycling model based upon tailored localized infrastructural configuration for all housing developments (see Figure 8.2). The Hammarby Sjöstad project has thus been constructed as an exemplar sustainability showcase within which a 'post-centralized network' ideal is intrinsically incorporated for 'green' water and energy service provision. What importantly distinguishes the Hammarby example (along with many similar projects) from Woking is its utility and (political) justification as a local 'showcase' for Stockholm as a whole, whereas Woking, at least originally, has been very much about developing a specific energy policy and solution for the whole borough.

While the alternative technological systems upon which such eco-districts in many urban regions are founded appear to echo modernist rationales of infrastructure as the harbinger of 'emancipatory futures', they have two major differences. First, they are developed on a much smaller geographical scale (between that of the building and of the neighbourhood), which may problematize issues of socio-spatial solidarities (see below). Second, they rest on a much more systemic, intersectoral approach. The Hammarby project, for example, is based wholly on the emblematic recycling model and rests on the systematic promotion of joined-up solutions for water, waste water, waste and energy (involving the Stockholm Water Company, the waste company, and energy providers Fortum and Fortum Värme together).

Beyond or before collective infrastructure

A third form of challenge to large networked systems can often be found at the edges or margins of cities in the North where traditional centralized infrastructure has not (yet) been extended, usually because of a combination of lower population densities, cost versus return on investment in network deployment, and technical difficulties in laying the necessary pipes or cables. This is particularly true for waste water and (non-electrical) energy systems, but also, in some cases, for water supply systems. These spaces beyond the network may be included in future network extension plans or be more permanently reliant on alternative forms of basic service provision which might be more adequate or pertinent in certain contexts. Many municipalities in the periphery of the Greater Stockholm region, for example, clearly distinguish in their planning strategies between zones already connected to municipal or inter-municipal water and waste water networks, those planned to be connected (in the short or longer term) and those that will remain permanently 'beyond' municipal networks (for a mixture of technical, geographical and economic reasons). These latter zones, where individual wells and septic tanks are prevalent, constitute quite a substantial part of the Stockholm archipelago region. In total, 100,000 households in the region are not connected to either or both water and waste water networks. In the largest municipality in terms of surface area, Norrtälje, fully 45 per cent of the population lives beyond the reach of centralized infrastructure networks.

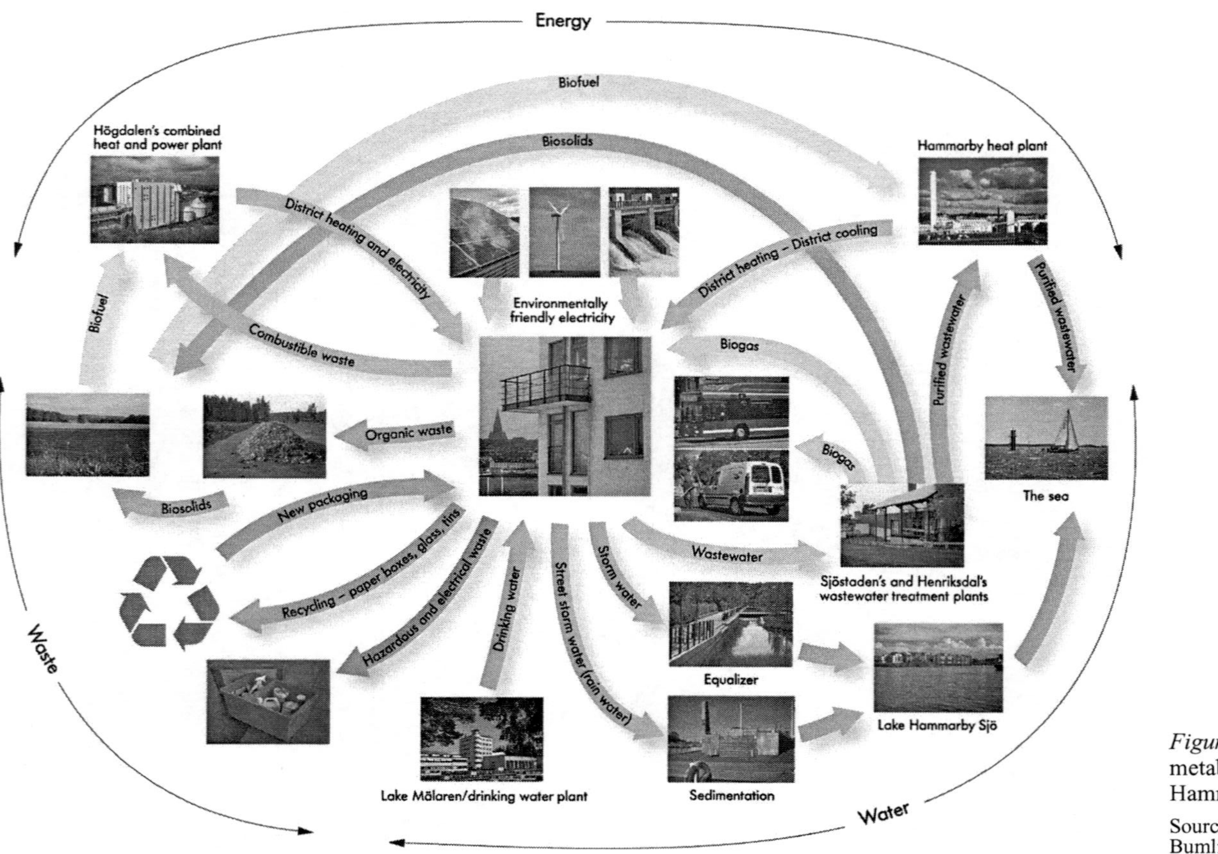

Figure 8.2 Closing the metabolic loop: the Hammarby model.

Source: Lena Wettrén, Bumling AB.

This approach, which assumes that less densely built areas will *not* be supplied by centralized networked systems on a permanent basis, opposes the still dominant approach whereby the extension of networks tends to follow, accompany or even anticipate the urbanization of new, suburban spaces. By no means recent in the Stockholm context, this approach may, so we argue, gain influence in many European cities or countries in the wake of increased critiques of centralized networks.

Feed-in to grid

The fourth form of 'post-networked' pathway to low carbon transition involves new forms of articulation between alternative or decentralized technologies and centralized systems. Policies introducing so-called feed-in tariffs to support the generation of energy from increasingly widespread decentralized techniques help to reinforce these new forms of organization of energy systems. These include individual power producers, including non-corporate ones. What is striking in these new arrangements is that independent power producers (IPPs) retain most or even full control over when they feed power into the grid, how much and for how long. Their input is not centrally monitored, as in traditional grid systems.

Hence, even though the network infrastructure remains crucial, the organization of the networked system is profoundly changed as centralized control is reduced and as the boundaries between power suppliers and consumers become increasingly blurred. This supports a splintering of the energy supply system, both as more and more IPPs emerge, including at the scale of individual buildings, establishing new relations of active interaction rather than the passive reception associated with the networked system, and as the notion of homogeneous, standardized supply to all consumers consequently wanes.

Urban significance and implications

We have outlined in the previous section diverse emerging forms of 'post-networked' urbanism for decentralized energy and urban-environmental service supply. These processes do not, however, reflect a move en masse to decentralized technologies. Nor do they imply that these decentralization trends go unchallenged. Politically, some authors have shown how, in France at least, the state was able to take advantage from the new European governance regime and economic reforms to reinforce its power *vis-à-vis* local governments (Poupeau 2004). Economically, 'liberalization' reforms have led to an increased, rather than a reduced, concentration in utility sectors (energy, telecommunications, transport, etc.). Technologically, in many instances 'decentralized' technologies have been integrated into centralized systems.

Nevertheless, we argue that these processes do constitute a combination of trends that challenge traditional, large networked systems as ideals and as material realities. They all involve the *scaling down* of systems of provision of energy and environmental services: from national to regional, from regional to community, from community to individualized systems. This downscaling and relocalization obviously runs the risk of complying with an unrealistic and normative conception of 'the local' as 'a socio-spatial container in which the sum of institutional, social and physical relations necessary to achieve a more sustainable future can be found . . . a "black box", disconnected from the global, international and national contexts within which localities are framed' (Marvin and Guy 1997: 312). We need, therefore, to better understand the variety of forms and ambivalent implications of the post-networked city. We argue that we might go some way towards this by focusing on the shifting patterns and tensions within and between the four dimensions of technical systems (introduced at the end of the first section) which any decentralized or 'relocalized' pathway to low carbon transition both implies and beckons.

Financing of urban service provision

Post-networked systems of provision of urban energy and environmental services imply changes to the ways in which these services are financed. Two distinct but concomitant processes are at stake here.

First, the overall cost of service provision is very likely to increase substantially, owing to rising environmental and quality expectations, regulations and standards, and to the duplication of infrastructure often involved in the development of decentralized technologies. Thus, although the 'off-grid' form has been promoted by public authorities in Woking, London, Southwark and others it has depended in these cases to a large extent on the mobilization of private finance. In Woking, the ESCO, albeit a public–private joint venture, invested in the CHP plant, heating and cooling system, and other local projects using the money of the private partner (which in turn gave it more financial freedom with respect to government regulations). In London, the LDA is 'bringing through' similar decentralized energy projects, but with a view to attracting long-term private investment in each case. Service provision over Southwark's new infrastructure will be financed by the new multi-utility services company (MUSCo), which involves Dalkia and Veolia. In return for its taking 100 per cent of the commercial risk, it will have a guaranteed customer base through the use of compulsory connection agreements in the Elephant and Castle project.

Second, major environmental service and energy companies such as Suez and EDF are having to drastically rethink their business models in a context of diminishing economies of scale, scope and club, rising total costs, changing revenue streams, and changing rationales of remuneration (from revenues based on the volumes sold to revenues based on the volumes or resources saved) – all of which increase the uncertainties over the economic and environmental 'sustainability' of service provision over traditional infrastructure networks (Saint-André 2008). According to an EDF official in London, the 'future proofing' of cities with 'smart', low carbon infrastructure will be inherently dependent on the capacity of utility companies to define realistic 'thirty-year business models' for investment in certain technologies (EDF Energy, London, interview, February 2009). In emerging business models based on energy services and the like, revenue generation is moving 'downstream' to the commercial part of the business. As a result, financial resources for the capital-intensive 'upstream' parts of the system (large power plants, grid infrastructure) are coming under increasing pressure.

Socio-spatial solidarities

Directly related to financing issues, post-networked city transitions potentially involve a recombination of the socio-spatial solidarities upon which most networked cities were constructed (and upon which many still operate). The introduction and promotion of decentralized technological systems, often under the auspices of environmentalism and the sustainability imperative, may well be at odds with the traditional wider social cohesion goals of network service provision, which tried to ensure equalities of access by some form of cross-subsidy system between user groups, areas and/or sectors. By downscaling the territorial extent of service provision, these systems and their users effectively withdraw from larger-scale, centralized systems and no longer contribute to their techno-economic feasibility. In the case of Hammarby Sjöstad, on one level the techno-ecological goals of the project for localized infrastructure configuration and a circular metabolism of resources and waste have steadily overrun the original social integration goals for a significant proportion of the rental apartments to be built to be accessible to low-income households. Hammarby Sjöstad is now packaged as a self-contained upmarket residential development. On another level, the metabolic loop-closing model implemented in Hammarby (and in similar projects around Europe and elsewhere) might be environmentally sustainable on a local level, but we can seriously question what the implications are for wider social solidarities on an urban scale if traditional integrated socio-technical systems, which depend intrinsically on widespread use and financial cross-subsidies, begin to fragment. In this regard, it might well be viewed as a clear-cut 'secessionary network space' geared to affluent middle-class families sensitive to ecological issues, thereby reinforcing trends of socio-spatial polarization in cities (Graham and Marvin 2001).

In contrast with the observed gentrification of the Hammarby Sjöstad neighbourhood, several projects carried by London boroughs seek to jointly attain social and environmental objectives through decentralized technologies. In Southwark (Elephant and Castle) and in Barking and Dagenham, for example, projects aim at providing existing deprived estates with 'affordable, low-carbon heat' from CHP systems, rather than creating an entirely new eco-neighbourhood. The resulting socio-spatial processes are likely to be very different, as 'ecological regeneration' in these instances does not rest on the extensive renovation and upgrading of the housing and building stock.

Urban metabolisms

Third, the rise of a post-networked city implies potential shifts in urban metabolisms and in patterns of resource circulation within (and without) the city. Both the 'off-grid' (e.g. Woking) and 'loop-closing' (e.g. Hammarby Sjöstad) forms involve the circulation of decarbonized energy, local production plants 'on site', the reuse of waste products and so on (see Figures 8.1 and 8.2). The scale of urban metabolism to which they contribute varies (local and translocal), but this relates to the changing economy of environmental services; what matters now (economically, socially, environmentally) is the flow more than the infrastructure, and the resource more than the flow. Implications of this change are illustrated, for example, by the double piping for potable water and non-potable water in the Elephant and Castle project in London, or by the development of district heating systems to use (most of) the heat generated by local combined heat and power plants. From another angle, it is also significant that, while Thameswey in Woking is focused on energy flows purely around Woking, the energy companies involved at Hammarby Sjöstad have more translocal interests which necessitate points of connection between the Hammarby circuit and wider electricity and heating networks; local metabolisms in this instance are subservient to broader evolving network economy logics.

However, relocalizing flows of water and waste water through the promotion of techniques such as wells, rainwater harvesting and grey-water recycling has been critiqued from an environmental health perspective (see, for example, Hardoy *et al.* 2001: 213 on urban services in developing countries). In a way, these debates echo hygienist arguments concerning urban proximity and densification in the eighteenth and nineteenth centuries (De Swaan 1995; Barles 1999). In the wider Stockholm region, there has been increasing concern about the environmental externalities of 'beyond the network' systems for water supply and waste water removal, involving salt water infiltration problems, insufficient separation of drinking water and waste water tanks, and pollution of the Baltic Sea from untreated waste water run-off. Likewise, waste incineration plants, which epitomize the new circular metabolism of 'sustainable' and 'low-' or 'post-carbon' cities, have been the object of fierce opposition by riparian populations and/or environmental associations (see, for example, Rogers 2005; Rocher 2008).

The governability of post-networked cities

The rise of post-networked urbanism also implies both reconfigured forms of infrastructure governance and, more broadly, changing relations between city authorities and their citizens concerning the provision of basic services. Dominique Lorrain (2003) has argued that cities have been historically governed first and foremost via the technical networks that irrigate and innervate them. This implies that any change in these networks is likely to have important repercussions for urban administration and governance.

The forms of post-networked urbanism outlined in the previous section suggest quite contrasting reconfigurations of governability. The 'off-grid' examples (Woking, London, Southwark) highlight local authorities' 'reseizing' control of infrastructures almost as an instrument of government (again) (for example as a way to provide cheaper, cleaner, secure energy to their citizens) or as a beacon for a more autonomous approach to low carbon socio-technical change. This approach implicitly (or explicitly) positions itself in opposition to the centralized grid system configuration, which, for local authorities such as Woking and Southwark, has, or had, a triple disadvantage of failing to provide inexpensive access to energy for households and businesses, of reinforcing dependence on fossil fuel-based energy supplies, and of making local authorities comparatively 'powerless' to shape change. In other words, the low carbon (and wider environmental) discourse favouring alternative, decentralized systems appears in some cases to have been mobilized also as a means for authorities to recover (at least partial) responsibility for basic service provision and the socio-technical infrastructures needed to achieve this. The 'loop-closing' context in Stockholm is different in that it has emerged in a context of 'demunicipalization' of service provision (the sale of Stockholm Energi to Fortum), but this is tempered by the continuing role of the municipality in the heating company and its wider coordination and planning role, which has succeeded in involving a private company (Fortum) in 'exemplary' environmental service provision objectives in neighbourhoods such as Hammarby Sjöstad. This has certainly contributed to maintaining the Stockholm municipality's strong environmental governance or leadership in the eyes of the local population (cf. Rutherford 2008). In contrast, the 'beyond the network' and 'feed-in to the grid' forms at the very edges of urban infrastructure governance are more reliant on private (household or utility) supply and generation. In the former case, in particular, authorities have taken a conscious decision that these sparsely populated, geographically isolated areas exceed the (financial/organizational/spatial) limits and possibilities of governability through collective infrastructure.

In sum, we argue that any undermining of the 'networked city' model is not, then, merely a technical issue of import only to those interested in the sub-discipline of urban infrastructure networks, but is central to the broader ongoing shift in the urban paradigm identified by other writers (Graham and Marvin 2001; Soja 2000; Amin and Thrift 2002; Dear and Flusty 1998). As Lorrain suggests, cities in the North and the South must cope permanently with 'a disjuncture between infrastructures, technical networks, institutions and real urbanization', which commonly leads to a spatial, social and functional 'gap' between 'an organized city and its fringes' (2003: 448–449). As is illustrated in the examples discussed above, there is no exact correspondence between the space of the city in which people live, the space of the city administered by its institutions, and the space of the networked city (which used to, and still does in many cases, provide the most direct link between local authorities and urban populations).

Which, and whose, post-networked city?

In this chapter, we have been concerned with exploring the distinct forms and ambivalent social implications of the post-networked urban-infrastructural turn as an emerging set of pathways towards low carbon transitions. These will not mechanically lead to utopian 'sustainable' cities that are more carbon neutral, more just, more liveable for all, but nor will they in themselves necessarily produce dystopian cities with socially and environmentally deprived areas next to secessionary green enclaves. Rather, as we have tried to illustrate, the social and urban implications of the development of decentralized technologies (DTs) in the provision of energy and environmental services will vary according to the socio-spatial context in which this development occurs and the mix of policies (especially those concerning land use and housing) with which it interacts.

Yet this socio-technical transition does potentially imply a more fundamental reconfiguration of the social, political, economic and environmental rationales and finalities not just of urban infrastructure but of cities themselves. Indeed, we have argued that the contradictions and tensions inherent to these developments lie at the interface between technological change and the multiple facets of 'the urban'. Thus, the variety of forms of post-networked urbanism are both constructed on and imply quite different financial, socio-spatial, metabolic and governance configurations. The governance-led socio-technical change of the 'off-grid' and 'loop-closing' forms contrasts with the more 'user'- or utility-oriented forms of 'beyond the network' and 'feed-in to grid'. Within each form, however, these arrangements are not fixed and stable, but remain highly contested and contradictory. The exemplary metabolic organization of Hammarby Sjöstad conflicts with the more problematic socio-spatial implications of this redevelopment area. The private financing and local authority-led 'off-grid' cases appear a little at odds with environmental and low carbon objectives, not to mention the network externalities susceptible to being produced by whole boroughs and municipalities withdrawing from centralized systems. The 'feed-in to grid' form has in some cases been criticized as 'a great green rip-off' in its transfer of money from higher energy bills to the few middle-class homeowners capable of installing solar panels on their roofs (Monbiot 2010). Indeed, we would argue that many 'post-networked' configurations may lead to the diversion of financial resources from the incumbent utility system (and, often, from the incumbent utility company as well) to the benefit of alternative local systems. Individuals and/or local communities may be increasingly reluctant to finance large systems perceived as 'unsustainable' and of declining importance in the provision of services. A plausible scenario would involve a vicious circle dynamic, in which diminishing financial resources for large infrastructure systems would concomitantly support and be supported by the declining quality and reliability of the services provided by those systems. A key prospective question thus concerns the financial sustainability of the incumbent infrastructure, which in most cases will remain of crucial importance, if only as a last-resort supply system, if we are to avoid a 'tragedy of the infrastructural commons'. A suspicion remains, therefore, that this transition towards a post-networked city is not always desirable, providing service 'niches' for a minority at the (economic, social, environmental) expense of the majority. As particular actors and spaces are relatively more excluded from or bypassed in the transition process, as their interests and agendas diverge from those of others, it is clear that they engage differently with and experience (or not) in varying forms and to varying degrees the spatio-temporal components and priorities of transitions. Place-based low carbon transitions such as those based on the promotion of DTs may thus be conceived as socio-political assemblages of the many diverse interpretations of the nature, form and spatial/temporal scales of change that actors develop according to their own (shifting) needs and agendas (see also Coutard and Rutherford forthcoming).

The ways in which post-networked urbanism (and the new techno-ecocycle paradigm that underpins it) can be adapted to fit the quite distinct conceptions, visions and requirements of proponents of both 'market sustainability' (the technologically innovative and competitive low carbon city) and 'strong sustainability' (the energy-sober, low-carbon-lifestyles city) merits reflection as to the work it does or implies. What accounts for this convenience and compatibility of DTs with regard to these policy positions? The introduction of decentralized technologies in many cases is justified by 'strong sustainability' rationales associated with climate change mitigation, urban resilience and autonomy and so on, which clearly demand radical change (as captured indeed by the notion of a post-networked city). Yet some of the forms of post-networked urbanism that we have highlighted rather seem to embrace substantially less radical ideas of ecological modernization with DTs conceived as merely a new, highly convenient techno-ecological fix for policy makers and utility companies to combine technological development, organizational practice and economic competitiveness. In spite of discourses, the development of DTs may often support rather than challenge the developmental, pro-growth (in terms of environmental resource use) logic associated with large networked systems. DTs are thus being enthusiastically deployed as the answer to sustainability requirements while broadly allowing 'business' (or consumption) as usual. They may in some cases be a means for policy makers and service suppliers to 'normalize' sustainability within their (business) plans or operations. They are often seen, as is the case with decentralized energy in London currently, as an emerging market potentially involving massive revenue streams (the same investors who backed start-up internet companies ten years ago are now backing new 'green' technology companies) and thus allow the supposed reconciliation of ecological preservation with economic growth, something that is particularly important in these times of global economic crisis. In sum, in spite of their interpretative and political flexibility, the diffusion of DTs and the concomitant (relative) demise of incumbent large technical networks will undoubtedly affect various social regulations in ways that will need to be well understood and properly addressed.

More progressive forms of 'post-networked city' will require more fundamental and integrated reconfigurations of urban ecologies, infrastructure and socio-spatial change. These more coalescent, systemic and inclusive understandings of urban transitions may promote sensitivity to the multiplicity and diversity of situations across, and especially within, different urban contexts. Sifting between and untangling the conflicting logics associated with the notion of a 'post-networked city' as at once a discursive infrastructural ideal, an analytical tool, a progressive alternative, a new techno-ecological (spatial) fix and a process reinforcing urban fragmentation demands further theoretical and empirical work. For all its connotations as a quasi-definitive outcome of contestations of traditional models of basic service provision, the 'post-networked city' is an emerging, systemic process, open to engagement and critique, and thus remains to be invented. It is nevertheless our contention that its initial contours and framework have at least some potential to reinvigorate study of the multitude of ways in which technology makes and remakes urban society within and beyond the current sustainable development agenda.

References

Amin, A. and Thrift, N. (2002) *Cities: Reimagining the Urban*, Cambridge: Polity Press.

Barles, S. (1999) *La ville délétère. Médecins et ingénieurs dans l'espace urbain XVIIIème–XIXème siècle*, Seyssel, France: Champ Vallon, coll. Milieux.

Coutard, O. (ed.) (1999) *The Governance of Large Technical Systems*, London: Routledge.

Coutard, O. (2010) 'Services urbains. La fin des grands réseaux?', in O. Coutard and J.-P. Lévy (eds) *Ecologies urbaines*, Paris: Economica-Anthropos.

Coutard, O. and Rutherford, J. (forthcoming) 'Energy transition and city-region planning: Understanding the spatial politics of systemic change', *Technology Analysis and Strategic Management* (in press).

De Swaan, A. (1995) *Sous l'aile protectrice de l'Etat*, Paris: Presses Universitaires de France.

Dear, M. and Flusty, S. (1998) 'Postmodern urbanism', *Annals of the Association of American Geographers* 88: 50–72.

Dupuy, G. (1985) *Systèmes, réseaux et territoires*, Paris: Presses de l'ENPC.

Elzen, B., Geels, F. W. and Green, K. (eds) (2004) *System Innovation and the Transition to Sustainability: Theory, Evidence and Policy*, Cheltenham, UK: Edward Elgar.

Geels, F. W. (2002) 'Technological transitions as evolutionary reconfiguration processes: A multi-level perspective and case study', *Research Policy* 31: 1257–1274.

Gilroy-Scott, B. (2007) 'How to get off the grid', in The Trapese Collective (ed.) *Do It Yourself: A Handbook for Changing Our World*, London: Pluto Press.

Graham, S. and Marvin, S. (2001) *Splintering Urbanism: Networked Infrastructures, Technological Mobilities and the Urban Condition*, London: Routledge.

Greenpeace (2005) *Decentralising Power: An Energy Revolution for the 21st Century*, London: Greenpeace.

Hardoy, J. E., Mitlin, D. and Satterthwaite, D. (2001) *Environmental Problems in an Urbanizing World*, London: Earthscan.

Harvey, D. (1985) *The Urbanization of Capital*, Oxford: Blackwell.

Hirsh, R. F. (2003) *Technology and Transformation in the American Electric Utility Industry*, Cambridge: Cambridge University Press.

Hodson, M. and Marvin, S. (2009) 'Urban ecological security: A new urban paradigm?', *International Journal of Urban and Regional Research* 33: 193–215.

Hughes, T. P. (1983) *Networks of Power: Electrification in Western Society 1880–1930*, Baltimore: Johns Hopkins University Press.

Joerges, B. (1999) 'High variability discourse in the history and sociology of large technical systems', in O. Coutard (ed.) *The Governance of Large Technical Systems*, London: Routledge.

Kaika, M. and Swyngedouw, E. (2000) 'Fetishizing the modern city: The phantasmagoria of urban technological networks', *International Journal of Urban and Regional Research* 24: 120–138.

Lorrain, D. (2003) 'Gouverner "dur-mou". Neuf très grandes métropoles', *Revue Française d'Administration Publique* 107: 447–454.

Marvin, S. and Guy, S. (1997) 'Creating myths rather than sustainability: The transition fallacies of the new localism', *Local Environment*, 2: 311–318.

Mayor of London (2007) *Action Today to Protect Tomorrow: The Mayor's Climate Change Action Plan*, London: Greater London Authority.

Mayor of London and Greenpeace (2006) *Powering London into the 21st Century*, London: Mayor of London and Greenpeace.

Melosi, M. V. (2000) *The Sanitary City: Urban Infrastructure in America from Colonial Times to the Present*, Baltimore: Johns Hopkins University Press.

Monbiot, G. (2010) 'Are we really going to let ourselves be duped into this solar panel rip-off?', *Guardian*, 1 March.

Montginoul, M. (2006) 'Les eaux alternatives à l'eau du réseau d'eau potable pour les ménages. Un état des lieux', *Ingénieries* 45: 49–62.

Newman, P., Beatley, T. and Boyer, H. (2009) *Resilient Cities: Responding to Peak Oil and Climate Change*, Washington, DC: Island Press.

Offner, J.-M. (2000) 'Territorial deregulation: Local authorities at risk from technical networks', *International Journal of Urban and Regional Research* 24: 165–182.

Patterson, W. (2005) 'Decentralizing networks', *Cogeneration and On-Site Power Production*, January–February: 21–25.

Poupeau, F.-M. (2004) *Le service public à la française face aux pouvoirs locaux. Les métamorphoses de l'Etat jacobin*, Paris: CNRS Editions.

Rocher, L. (2008) 'Les contradictions de la gestion intégrée des déchets urbains. L'incinération entre valorisation énergétique et refus social', *Flux* 74: 22–29.

Rogers, H. (2005) *Gone Tomorrow: The Hidden Life of Garbage*, New York: The New Press.

Rotmans, J., Kemp, R. and van Asselt, M. (2001) 'More evolution than revolution: Transition management in public policy', *Foresight* 3: 15–31.

Rutherford, J. (2008) 'Unbundling Stockholm: The networks, planning and social welfare nexus beyond the unitary city', *Geoforum* 39: 1871–1883.

Saint-André, B. (2008) 'L'énergie intelligente dans la ville durable. Perspectives d'évolution du métier d'énergéticien', *Flux* 74: 68–76.

Shove, E. and Walker, G. (2007) 'CAUTION! Transitions ahead: Politics, practice, and sustainable transition management', *Environment and Planning A* 39: 763–770.

Smith, A., Stirling, A. and Berkhout, F. (2005) 'The governance of sustainable socio-technical transitions', *Research Policy* 34: 1491–1510.

Soja, E. (2000) *Postmetropolis: Critical Studies of Cities and Regions*, Oxford: Blackwell.

South West Renewable Energy Agency (2007) Woking Borough Council Energy Services Company. Online, available at: http://domain1723054.sites.fasthosts.com/downloads/RegenSW_99.pdf (accessed 20 January 2009).

Summerton, J. (ed.) (1994) *Changing Large Technical Systems*, Boulder, CO: Westview Press.

Tarr, J. A. (1997) *The Search for the Ultimate Sink: Urban Pollution in Historical Perspective*, Akron, OH: University of Akron Press.

Tarr, J. A. and Dupuy, G. (eds) (1988) *Technology and the Rise of the Networked City in Europe and North America*, Philadelphia: Temple University Press.

Willis, R. (2006) *Grid 2.0: The Next Generation*, London: Green Alliance.

9 Living laboratories for sustainability

Exploring the politics and epistemology of urban transition

James Evans and Andrew Karvonen

Introduction

Creating a more sustainable society is increasingly an urban challenge (Pincetl 2010). Upwards of 50 per cent of the world's population currently dwell in cities, and this figure is forecast to rise dramatically over the coming decades (Grimm *et al.* 2008). Cities both concentrate the activities that produce carbon emissions, and suffer disproportionately from their negative impacts such as air pollution, temperature increases, water shortages and increased flooding. Given this, cities are increasingly being looked to as sites to develop long-lasting solutions to climate change (Hodson and Marvin 2007).

This chapter focuses on the use of 'living laboratories' to drive innovation in sustainable urban development. The types of spaces designated as living laboratories are highly variable, from a single plot of underdeveloped land to a degraded waterway, from a clogged transportation corridor to a completely new city. Further, a wide variety of organizations – notably universities, government bodies and private companies – are using the term in an unapologetically boosterish manner to develop and market their own approaches to sustainability. Their enthusiasm is underpinned by two assumptions: first, that living laboratories are real-life experiments that promise to produce more *useful* knowledge; and second, that they are highly *visible* interventions with the purported ability to inspire rapid social and technical transformation. Taking a series of examples, we consider the epistemological and political implications of living laboratories, asking whether such experiments really do hold the potential seeds of change, as this literature suggests, or whether there are other motivations at work. The chapter concludes with a discussion of role of the living laboratories approach as a form of experimentation in relation to theories of transition and sustainable urban development.

Laboratories and knowledge generation

The word 'laboratory' implies an epistemology; it is a way to know the world. Laboratory studies scholars have furnished well-known accounts concerning the exceptional qualities of laboratories as privileged spaces that channel and accentuate the power of science (e.g. Latour and Woolgar 1979; Knorr-Cetina 1981; Lynch 1985). Because they are purposefully separated from the lived world, they allow variables to be isolated and carefully manipulated in order to test hypotheses (Knorr-Cetina 1995). The material practices that take place in laboratories involve the enculturation of natural objects, breaking them down into constituents that can be examined under conditions dictated by the exacting demands of the experiment rather than the unpredictable whims of nature. In this sense, 'laboratories create enhanced environments where it becomes possible to see things not visible elsewhere' (Henke and Gieryn 2008: 362). Constructed through specific scientific practices, laboratories inscribe accounts of reality that can be repeated dependably, transforming the findings of an experiment in one place at one time into placeless facts that can be transported anywhere while remaining unchanged (Latour 1987).

The authority of laboratory knowledge depends upon this placelessness (Kohler 2002). As Henke and Gieryn (2008: 353) write, 'the laws of gravity worked the same everywhere; even if scientists in different locations disagreed for a time about the content of these laws, persuasive evidence and compelling theory would eventually rub out geographical differences in belief'. Part of this 'rubbing out' involves the generation of a social order to go alongside this new objective order, as the constructors (in the traditional account, scientists) are reconfigured around the new 'facts' (Knorr-Cetina 1995). Latour (1998) argues that this specific quality of laboratories allows them literally to 'raise the world' that we know into existence, remaking objects and subjects (facts and society) simultaneously.

But if the gold standard for natural scientists is knowledge produced in carefully isolated conditions, then how are we to understand 'living laboratories', which claim to produce laboratory ways of knowing in the notoriously messy real world? After all, it is precisely this messiness that, historically, has led scientists to denigrate field-based methods as vastly inferior to the control and explanatory power offered by laboratories. On this understanding, 'living laboratories' should be consigned to the epistemological trash can; at best a contradiction in terms, and at worst a pathology of pseudo-scientific posturing. But the relationship between the lab and the field is often more permeable than the traditional account acknowledges; indeed, the very idea that the 'lab' can be separated from 'reality' has been widely critiqued (see Bowker 1994; Kuklick and Kohler 1996; Henke 2000; Gieryn 2006). As such, the living laboratory approach offers promise for redefining what it means to experiment and innovate in the remaking of the world.

Kohler (2002) defines the lab–field dichotomy as a border zone in his historical account of US biological studies. He follows the attempts of successive researchers to reconcile the supposed superiority of lab methods with the necessity of working in the field, which classic problems like speciation, by their nature, required. Practices of place play a central role in his account. Early field biologists sought out places in which unusual circumstances had created natural experiments, allowing them to mimic the control of a lab. Charles Darwin famously called the Galapagos a living laboratory for the study of evolution because its unique geographical isolation replicated the variables of separation required to study speciation. But these studies largely failed in their efforts to empirically verify speciation, leading the pioneers of population biology in the 1920s and 1930s to turn this practice of place on its head. Rather than identifying special or unique settings to study nature's experiments unfolding, site selection was driven by the practicalities of collecting large amounts of data, which, in turn, privileged ease of access. The paradigmatic example is Raymond Lindeman's field study of Cedar Creek Bog in Minnesota, which yielded the trophic-dynamic theory of energy flow that forms the conceptual bedrock of modern ecology: ecosystems. Cedar Creek was chosen because of the ease with which data could be collected, allowing researchers to reveal the interworkings of its ecosystem dynamics inexpensively; it had a very simple species structure, and was shallow, so could be cored to reveal species compositions over many years. Population biology managed to develop explanatory analyses from field studies by collecting an exhaustive amount of data, so that it became possible to identify and isolate variables and test causal links within. In this way, techniques of the field were developed to fit the rules of knowledge production as practised in the lab.

Living laboratories, innovation and transition

The concept of the 'living laboratory' blurs the distinctions between laboratory and field, inside and outside, controlled and uncontrolled experiment in a similar way to Kohler's population biologists, and his account has relevance to the way in which 'living laboratories' are being used to study sustainability today. The systems studies literature presents the living lab approach as a research methodology for sensing, testing and refining complex solutions in real-life contexts, a methodology that is user driven and focused on generating innovation through public–private partnerships (Niitamo *et al.* 2006). The most prominent contemporary examples of living labs can be found in the field of commercial product development. CoreLabs, a coalition of high-profile European academic and private organizations, provides a typical definition of living labs as 'functional regions' where stakeholders (including firms, public agencies, universities, institutes and individuals) form a public–private partnership to collaboratively create, prototype, validate and test new services, products and systems in real-life contexts. The phrase 'functional region' refers to cities, villages, rural areas and industrial plants, thereby separating the living laboratory *in* the real world, if not *from* it. In a familiar refrain, the company goes on to state that living labs are superior to 'closed' labs 'in virtually all aspects' because they focus on user communities embedded within 'real life' (CoreLabs 2010).

Living labs are notable for their reliance on cutting-edge information and communication technologies (ICT) to realize innovation (e.g. Abowd *et al.* 2000; Frissen and van Lieshout 2004). In a widely cited paper, Kidd *et al.* (1999: 193) describe the development of an 'Aware Home' as a living lab for the study of domestic activity. The purpose-built house was fitted with ultrasonic sensors, radio-frequency and video technologies, floor sensors and vision techniques, all of which feed data back to the domestic computer which responds to the needs of the inhabitants. Mirroring Kohler's field biologists of the early twentieth century, Kidd *et al.* (ibid.: 191) describe the aware home as an 'authentic yet experimental setting' – a living lab that offers both the control of laboratory experimentation *and* the authenticity of real life. The access offered by the living lab allows ubiquitous computer-assisted learning to generate massive amounts of data from everyday activities. But the Aware Home takes the ease of access of Lindeman's bog to the next level, creating a bespoke space hardwired with monitoring equipment to capture live data. This commitment to total data capture runs through the literature, and is not confined to ICT monitoring. Multiple methods, ranging from creativity groups and social network logging to field trials and participatory design, are advocated as ways to collect data on all aspects of living labs.

Surprisingly, the centrality of the human user in the living labs approach unwittingly addresses a burgeoning critique from within the critical geography and Science and Technology Studies literature, which argues that sustainability transition is being framed in an overly technocentric way (see Brand 2005; Oudshoorn and Pinch 2005). By contrast, living labs emphasize the human dimensions of new technology uptake, arguing that it is users who ultimately determine the success or failure of innovations. Living labs thus validate products and services in collaborative, multicontextual, empirical, real-world environments, integrating users and stakeholders in the activities of product and service development. The individual here is understood as multifaceted and performing multiple roles, from user and customer to worker and maintainer (Kusiak 2007). Returning to Knorr-Cetina's language, the living lab approach interpolates technology into society at the most basic level, as co-evolving entities. It harnesses the power of the laboratory to generate new social conditions through commercial innovation.

The idea that innovation is the primary source of economic success has a long history (Schumpeter 1934), but studies of innovation have failed to achieve anything approaching a scientific insight into the process. Numerous disciplines have their own terminology to describe similar elements such as degree of novelty (incremental versus radical), types of innovation (products, systems, processes and services), and so forth, but two requirements tend to be widely accepted: first, innovation must create *new knowledge* (e.g. Brockman and Morgan 2003); and second, it must generate *commercial success* (e.g. Galanakis 2006). Procter & Gamble, the international consumer goods corporation, is a well-known example of the success of this approach and a leader in research innovation. The company has realized significant gains since including stakeholders in its living lab activities, doubling its innovation success rate in two years with no increase in research and development expenditure (Kusiak 2007).

The commercial living lab approach is experimental in a colloquial sense: many small innovations can be trialled at a low level (i.e. letting a thousand flowers bloom), which are then selected through an evolution of ideas. In this so-called innovation pyramid, competition and adaptation between products leads to the emergence of successful innovations (Utterback and Abernathy 1975). Geels's

(2002) model of socio-technical transition relies on a similar evolutionary approach to understanding how isolated innovations, occurring in specialized or protected 'niches', either are or are not scaled up through wider adoption. Within Geels's framework (see Chapter 2, this volume), niches can force change at the regime level of wider governance if they are suitably successful, either by scaling up or multiplying and eventually coalescing. As Geels notes, 'while regimes usually generate incremental innovations, radical innovations are generated in *niches*' (2002: 1260; emphasis original).

Living labs constitute classic niches for innovation in this sense, as arrangements 'built between actors to support the innovation in very specific time and space contexts' (Beveridge and Guy 2005: 675) that shelter it from wider political and economic pressures. That said, they differ from the classic examples offered by Geels in that they are *explicitly* experimental, rather than rationalized *post facto* as innovative. So, for example, the evolution of steam in the shipping industry and the developments in logistics that accompanied it were innovations (Geels 2002), but not in the same way that a living lab sets out explicitly to *create* a niche in which experimentation can occur.

Sustainable urban innovation at universities

Living labs are highly appealing to the wide range of actors involved in climate-change mitigation and adaptation activities. Rooted in a specific place, they offer the immediate real-world relevance sought by policy makers; data-rich, they offer the promise of causal understanding and 'factual' knowledge that is the *sine qua non* for scientists; user led in character, they generate innovations that have a greater likelihood of successful adoption, promising commercial rewards for the private sector. Further, the interdisciplinary approach of living labs lends itself to the academic challenge of urban climate adaptation, which Ravetz (2009) suggests requires innovation simultaneously in physical and social science, and a policy environment that increasingly speaks the language of demonstrations, experiments and pilot projects.

While the conventional commercial living laboratory model is not automatically associated with universities, living labs for sustainability frequently involve partnerships that include universities, the public sector and private companies. Living laboratories form part of the broader shift towards more pragmatic epistemologies that accompanies increasing pressure for universities to pursue research agendas that have relevance for society at large (Barnett 2000; Perry 2006). While notoriously hard to achieve in the urban environmental sphere (Evans 2006; Evans and Marvin 2006), an emphasis on demonstrable utility within academia more widely is set to continue, with the United Kingdom's forthcoming Research Excellence Framework requiring researchers to explicitly consider their impact on society (see Demeritt 2010). Table 9.1 includes a summary of four case studies of university-led living laboratories for sustainability that demonstrate four distinct approaches, each focusing on a different type of innovation which, in turn, privilege different actors and forms of knowledge. In the following section, we

Table 9.1 Four university-led living laboratory approaches

	Masdar City	*North Desert Village*	*Oxford Road Corridor*	*Urban Landscape Lab*
Driver of innovation	Technological breakthrough	Scientific knowledge	Collaborative urban governance	Socio-material adaptation
Dominant disciplines	Engineering, business	Ecology, biology	Public policy, engineering, business	Landscape architecture, architecture, planning, urban design
Fabricated?	Yes	Yes	No	No
Types of data collected	Environmental and social use	Environmental and stakeholder preferences	Environmental	Stakeholder knowledge and opinions
ICT use	Heavy	Heavy	Selective	Low
Scale	Large	Medium	Large	Small
Commercial emphasis	High	Low	High	Low

tease out their central characteristics to inform the ensuing discussion about living labs and transition.

Masdar City, United Arab Emirates

Masdar City is an ambitious project to construct a city of 50,000 residents 17 kilometres from Abu Dhabi in the United Arab Emirates. Designed by Foster & Partners as a model of sustainable urban development, it promises to be a zero carbon, zero waste city powered entirely by renewable energy (Masdar City 2010). The tagline for the project boasts that 'one day all cities will be like this', and Masdar CEO Dr Sultan Ahmed Al Jaber states, 'Masdar City is a blueprint for cities around the world striving for sustainability, and a living laboratory to advance renewable energy solutions' (quoted in Masdar 2010). Prosaically enough, the Arabic word *masdar* means 'source', and the strategic goal of the project is to generate ideas and knowledge that will make Masdar City the model for sustainable urban development throughout the world. It is arguably the most enterprising form of living laboratory to tackle the issue of climate change, and does so by forwarding technological development as key to a carbon-free future. Reflecting on the ambitious goals of the Masdar project, a journalist writes, 'Abu Dhabians are betting that technology can dominate the climate, and, with almost limitless resources, they just might succeed' (Hartman 2010).

The new university is a key aspect of this technopole of climate change innovation. The Masdar Institute of Science and Technology, unveiled in September 2009, was developed in collaboration with the Massachusetts Institute of

Technology (MIT). By 2011, the institute will offer ten postgraduate science and engineering programmes in renewable energy and sustainable technologies, and eventually it will be home to 600 graduate students and over 100 faculty to serve as a hub for public and private research institutes around the world. In addition to conducting traditional research and development on renewable energy technologies, the institute will be housed in a building that will be wired with an energy-metering system to monitor energy consumption and be used by the faculty and students as a research tool (see Masdar 2010).

Masdar is a large-scale, technology-dominated approach to urban transition that seeks to develop commercially viable solutions to urban sustainability in a fabricated living laboratory. The laboratory concept has at least two meanings in Masdar City. First, the city itself is a test-bed for a carbon-free lifestyle. A hybrid of vernacular Arabic and cutting-edge energy efficiency strategies are being incorporated to create a real-life city with a negligible impact on climate change. Second, the laboratory term refers to the development of a clean-tech sector, where the 'aim is to become the silicon valley for clean, green and alternative energy', with a high concentration of global and start-up companies that will use the city as a base for an energy revolution (Masdar City 2010). Innovation here is fuelled by technological breakthroughs and the increasing business opportunities offered by globalized energy markets.

North Desert Village, Arizona, USA

A second example of the university-led living laboratories approach is the North Desert Village in Phoenix, Arizona. Funded by the US National Science Foundation (NSF), the North Desert Village is a project within the Long-Term Ecological Research (LTER) programme, the flagship environmental science research programme in the United States, comprising twenty-four ecologically diverse sites, an annual direct budget of almost $20 million and approximately 1,100 scientists and students. In 1997, the NSF added two metropolises (Phoenix and Baltimore) to its portfolio of sites, and both projects were granted second-phase funding in 2004 (NSF 2002). The research teams in both cities adopted a large-scale ecosystems approach while also framing their cities as living laboratories (Grimm and Redman 2004). The Central Arizona–Phoenix LTER has realized this goal most explicitly by establishing an entire 'experimental suburb' to develop innovative urban ecological approaches.

Like Masdar, North Desert Village is entirely fabricated and is the first-ever experimental study of interactions between people and their ecological environment at the neighbourhood scale. One type of experiment involves the manipulation of vegetation types and irrigation methods to explore how landscape interactions affect human perception and behaviour. Residential landscapes at identical housing units in the village were installed in four different styles based on the different habitats found throughout the Phoenix area, ranging from a mixture of exotic water-intensive vegetation and shade trees with turf grass maintained by flood irrigation (reproducing the classic and largely unsustainable suburban garden type) to native plants on granite substrate with no supplemental water whatsoever (reproducing the surrounding Sonoran Desert landscape).

Faithfully reproducing the mantra of living laboratories, the researchers claim that experimenting on humans *in situ* produces 'more accurate scientific models' which are better able to capture the complex and unpredictable feedback mechanisms between social and ecological systems that typify urban areas (Cook *et al.* 2004: 467). The preferences of North Desert Village residents were used to design the area, and subsequently to inform ecological management decisions. This approach, dubbed 'adaptive experimentation', takes the user-centred approach of living laboratories to its logical conclusion, allowing users to adapt the experiment and alter its parameters from within. But however radical this methodology is, it is accompanied by a need to legitimize the North Desert Village as a scientific space. The experiment team does this through its description as a laboratory that incorporates 'most of the formal aspects of classic experimental design, including independence of study units, use of replicates, and controls' (ibid.: 467). The team's mode of adaptive experimentation turns the neighbourhood into a living laboratory, not by hardwiring science into the direct administration, design and planning of the urban landscape, but by assimilating these functions into the study itself (Evans 2009). Through this systems approach, nothing remains outside the lab. Unlike in Masdar, the innovation is decidedly scientific in character, promising to integrate environmental monitoring and social preferences in a sustainable urban management system. Like Masdar, the North Desert Village experiment includes user preferences, but the interface between users and landscape is never direct; it is mediated though a data monitoring and policy/management interface system. As such, this living lab is defined by technocentric forms of user participation that fit within the ecological science model of the experimenters.

Oxford Road Corridor, Manchester, UK

Not all living laboratories are created anew. Manchester's Oxford Road Corridor is an example of how the living labs approach can be superimposed upon an existing urban area. Comprising 243 hectares, two universities and five hospitals, the Oxford Road Corridor is frequently touted as 'the backbone of the city's knowledge economy', involving 37,000 employees, which represents 12 per cent of the city centre's workforce, and has an estimated annual turnover of £3.2 billion (MSCP 2008: 2). In the mid-2000s, the public and private stakeholders along the corridor partnered with the Manchester City Council to create the Manchester City South Partnership (MSCP), whose mission is to improve the economic, environmental and social aspects of the corridor, with the hope of generating spill-over benefits for the city, region and the United Kingdom. The partnership is committed to spending an estimated £2.5 billion over the next two decades on economic development, cutting-edge communication networks, an integrated mass transportation plan, green infrastructure and cultural growth. The University of Manchester wants to use the corridor, referred to by various actors as the 'Green Laboratory', 'Corridor Manchester' and 'Laboratory Manchester', to realize an affordable and resilient low carbon economy. Similarly to Masdar, the university intends to use Greater Metropolitan Manchester as a 'test-bed' for energy, communications and transportation technologies (Ravetz 2009: 8), with the corridor as the first major intervention. Both Masdar and the Oxford Road Corridor have a strong emphasis on commercially driven innovation that will eventually recoup the costs of building and running these large, real-world experiments.

From an urban development perspective, the university is involved as a large property owner that will be directly affected by any changes to the corridor. But there are increasing calls from both inside and outside actors for the university to become involved as a research partner in the initiative. In 2009, the University of Manchester and Manchester City Council signed a Memorandum of Understanding (MOU) to formally pursue climate change adaptation best-practices methodologies for urban environments. One of the first partnerships of this MOU is the EcoCities project, which will create a blueprint for climate adaptation in Greater Manchester by 2011 (see EcoCities 2010). Moreover, there are multiple ongoing and proposed research projects to monitor environmental parameters of the corridor (temperature, wind, air pollution, precipitation, etc.) using temporary and permanent monitoring climatic stations.

Like Masdar City, the living laboratory of the Oxford Road Corridor is firmly rooted in natural sciences and engineering, but the overarching emphasis here is not on technological innovation but rather on environmental policy and urban development. Innovation is achieved through new modes of governing the carbon flows in Greater Manchester under a rubric of ecological modernization that promotes a win–win scenario between economic and environmental interests, an aim that fits in with Manchester's designation as a Low Carbon Economic Area as well as the city council's Climate Change Action Plan (both unveiled in December 2009). The corridor serves as the pilot project for an overarching climate change agenda that will eventually shape the Greater Manchester metropolitan region as a whole.

Urban Landscape Lab, New York, USA

A fourth example of a university-led living laboratory comes from the design disciplines, notably architecture, urban planning, urban design and landscape architecture. Here, urban sustainability is frequently addressed through participatory design studios that focus on real-world issues. Such programmes contribute to a larger trend of project-based education where university research is directed to concrete problems to fulfil the university's commitment to public service (see Bajgier *et al.* 1991; Pearson 2002; Moore and Karvonen 2008). An exemplar of this approach is the Urban Landscape Lab (ULL) at Columbia University's Graduate School of Architecture, Planning, and Preservation in New York City. The laboratory is directed by Kate Orff and Janette Kim, two educators and designers, who lead an interdisciplinary, applied research group 'dedicated to advancing dialogue and new design methodologies to effect positive change in urban ecosystems' (ULL 2010). Their projects bring together a wide range of design and professional disciplines – architecture, landscape architecture, urban design, preservation, civil engineering, conservation biology, economics, climate, public health, and community-based advocacy – to work on traditional academic research and design studios as well as outreach activities that interpret 'cities as pedagogy' (Shields 2008: 714).

An example of a ULL project specifically directed at climate change adaptation is 'New Natures: Visualizing CarboNYC', a summer 2008 studio that visualized future scenarios for a self-sustaining New York City. The project examined settlement patterns at three different sites to understand how changing material conditions could be addressed with innovative urban design interventions. In this way, living laboratory experiments are used to rework the relationship between urban residents and their material surroundings. The scale of these design interventions tends to be smaller, owing to time and personnel constraints, with site and neighbourhood scales being the most common focus. There is an explicit recognition that the city is not an economic engine, a disrupted ecosystem or a policy conundrum, but rather a place of lived experience. And it is here, at the human scale, that sustainability interventions are considered to have the greatest impact.

Like the Oxford Road Corridor, the Urban Landscape Laboratory is not afforded a *tabula rasa* upon which new urban development strategy can be tested. Instead, the experimenters must contend with the existing fabric of the city, which includes a whole host of physical, economic, cultural and social relations. But while the Oxford Road Corridor focuses on top-down policy integration, to be achieved through collaborative urban planning and policy intermediaries, the ULL takes a bottom-up approach to socio-material adaptation of the urban environment by civic designers (e.g. Borasi and Zardini 2008). The living labs that have been fabricated (i.e. specially built) tend to be more comprehensively data driven, with a higher reliance on ICT and a notion of their user community similar to that found in the living labs literature. This reflects the ease with which data monitoring equipment and protocols can be incorporated into an environment that has been specifically constructed. The Oxford Road Corridor and ULL are inevitably confronted with greater data collection difficulties as they attempt to establish living laboratories in already existing facets of the urban environment. This difficulty may explain their emphasis upon human agency as a driver of change, rather than the technical-expert solutions that drive fabricated labs.

Transition and the politics of the living laboratory

In his study of the Clark Center biotech lab at Stanford, Gieryn (2008: 797) argues that the lab itself is an 'experiment' in linking the production of scientific knowledge with economic interests, adopting a design aesthetic that mirrors the disembedded architecture of high capitalism associated more commonly with airports and art galleries. The living laboratory for sustainability consummates a similar union between truth and economy. The implication of Masdar City and North Desert Village is that if they are deemed successful, then the knowledge generated there has 'added value' and is, in some sense, *more* valid than knowledge produced either elsewhere or in 'traditional' scientific settings. Places like Masdar and Oxford Road echo Henke and Gieryn's (2008: 365) observation that 'spaces for science are a powerful blend of material infrastructure and cultural iconography that lend credibility to knowledge claims'. As truth becomes synonymous with commercial success, it is embedded in the city as high-profile living laboratories in no less material a way than monuments to economic success, such as skyscrapers.

For sustainable urban development, living laboratories exist as truth spots – visible arbiters of truth in their own right that have their own logic. They are symbols of power and hold the promise for improved urban futures, but at the same time they are not simply blueprints that can be 'rolled out' everywhere. Rather, it may be the container itself that has viral properties. Kohler (2008) suggests that labs display an exceptional weediness of character, which allows them to permeate life in multiple ways and contexts. In terms of their ubiquity and plasticity, living labs may be part of the next stage of this dispersal. In his famous account of the pasteurization of France, Latour (1988) describes how the proliferation of labs in universities, commercial research facilities, farms and processing factories changed procedures and material realities across the country, interpolating the microbe into French society as much as they did into the natural order. In this way, the world became colonized by 'extramural' laboratories (Knorr-Cetina 1995: 160), which were outside but under the aegis of the scientific sphere. The concept of extra-mural colonization perfectly captures the process enacted by living labs, which explicitly endeavour to reproduce scientific ways of knowing in the real world. Further, the power of labs to interpolate seems more potent in relation to the applied scientific knowledge or techniques demanded by sustainability, whereby a range of users, builders, recorders and so forth are involved in the production of new knowledge. Living labs for sustainability interfere quite purposefully, harnessing the power of laboratories to remake society in accordance with new forms of knowledge.

Reflecting on Geels's framework of technological transition, living laboratories can be understood as a particular type of niche that explicitly incorporates knowledge production into decision making. Geels's model is formal in the sense that the content of the niche is inconsequential; it simply describes how content spreads when successful, eventually being adopted more broadly in the governing regime. Living labs constitute a special kind of niche that is knowledge driven, socially aware and *explicitly* experimental, qualities that are missing in Geels's framework (see Geels and Schot 2007 for a discussion). This self-awareness is crucial to the living laboratories approach because it opens up multiple avenues for, and a politics of, sustainable urban transition.

Living laboratories intentionally sidestep the tensions between bottom-up and top-down approaches to innovation in favour of lateral partnerships. On the one hand, this can be interpreted as a classic symptom of the post-political malaise affecting sustainability, where environmental catastrophe presents an opportunity for neo-liberal actors to wrest control of urban development from the government (see Swyngedouw 2007, 2009). On this understanding, climate change becomes a business opportunity for private interests to gain market share in the guise of helping society at large. Living laboratories aid and abet this appropriation by transforming the real world into a *de facto* research and development facility for capitalist modes of ecological modernization. The post-political landscape that emerges is one of the city as a 'self-adapting' landscape, controlled by the market and governed by the integrated monitoring and management of feedback loops between social and physical processes, in which political choices are tamed and replaced by user patterns and preferences (Evans 2009).

On another reading, though, the living laboratory approach offers the potential for opening up climate mitigation and adaptation strategies to new strategies and ideas informing more democratic modes of governance that are socially inclusive and sensitive to local conditions (Evans *et al.* 2009). This is precisely the trajectory taken with the new World Urban Campaign launched by UN-Habitat in 2010 through its abandonment of the 'best practices' approach of sustainable urban development in favour of a 'living practices' approach that is bottom-up, networked and intended to champion the unique lived experiences of individuals in particular cities (Peirce 2010; see also Bulkeley 2006). From this perspective, living laboratories have the potential to act as intermediary spaces positioned between technological possibilities and local contexts to 'manage' processes and governance of transitions (Hodson and Marvin 2009: 522) at the increasingly important levels of the city and region (e.g. Bulkeley and Betsill 2005; While *et al.* 2009). Here, climate change innovation emphasizes new modes of politics that challenge the conventional structural approaches that are top-down or bottom-up. Living laboratories are sites where 'fresh politics as yet unrecognized as such are emerging' (Latour 1998: 268); but precisely what form this politics will take is still up for grabs.

Conclusions

Kohler (2002: 214) notes how the idea of the natural laboratory formed a powerful part of field biologists' 'imaginative infrastructure' through which they understood the process of experimentation. Today, the rhetoric of living labs performs a similar function for those pursuing sustainable urban development. Hugely powerful yet poorly defined, living labs offer a set of alluring promises: as idea factories for generating relevant and usable knowledge, as 'test-beds' for applying this knowledge in real-world situations and as places to form the 'blueprint' for climate change mitigation adaptation elsewhere. Living labs have *already* had a major impact, as cities and universities rush to establish their own bigger and better versions to compete with those that already exist. However, it is uncertain whether these niche experiments can be scaled up to generate widespread changes in the existing processes of urban development. This depends in large part on how the knowledge produced through experiments is packaged and transferred to other locales. While all living labs express a desire to influence the wider world, the exact strategy or mechanism with which to scale up these activities is rarely outlined in detail; it is simply assumed that those successful innovative practices will somehow infiltrate and propagate, making these innovative practices the norm.

On the one hand, living labs (and experimentation in climate governance more widely) simply reproduce capitalism, seeking to 'tame' the messy realities of cities. On the other, the urban setting contains the seeds of many possible experiments that can resist this structural analysis; there is nothing about the urban condition that prefigures certain kinds of experiment. Indeed, explicit experimentation of the kind being practised in living labs can accommodate a wide range of collaborative partnerships, and it is perhaps this new mode of knowledge generation, rather than the knowledge created therein, that makes living laboratories such an enticing concept for realizing improved urban futures.

References

Abowd, G. D., Atkeson, C. G., Bobick, A. F., Essa, I. A., MacIntyre, B., Mynatt, E. D. and Starner, T. E. (2000) 'Living laboratories: The Future Computing Environments Group at the Georgia Institute of Technology', *Extended Abstracts of the ACM Conference on Human Factors in Computing Systems*. Online, available at: www.cc.gatech.edu/fce/ahri/publications/livinglabs-chi2000.pdf (accessed 15 January 2010).

Bajgier, S. M., Maragah, H. D., Saccucci, M. S., Verzilli, A. and Prybutok, V. R. (1991) 'Introducing students to community operations research by using a city neighborhood as a living laboratory', *Operations Research* 39: 701–709.

Barnett, R. (2000) *Realizing the University in an Age of Supercomplexity,* Buckingham: Society for Research into Higher Education and Open University Press.

Beveridge, R. and Guy, S. (2005) 'The rise of the eco-preneur and the messy world of environmental innovation', *Local Environment* 10: 665–676.

Borasi, G. and Zardini, M. (eds) (2008) *Actions: What You Can Do with the City*, Montreal: Canadian Centre for Architecture.

Bowker, G. (1994) *Science on the Run: Information Management and Industrial Geophysics at Schlumberger, 1920–1940*, Cambridge, MA: MIT Press.

Brand, R. (2005) *Synchronizing Science and Technology with Human Behaviour*, London: Earthscan.

Brockman, B. K. and Morgan, R. M. (2003) 'The role of existing knowledge in new product innovativeness and performance', *Decision Sciences* 34: 385–419.

Bulkeley, H. (2006) 'Urban sustainability: Learning from best practice?', *Environment and Planning A* 38: 1029–1044.

Bulkeley, H. and Betsill, M. (2005) *Cities and Climate Change: Urban Sustainability and Global Environmental Governance*, London: Routledge.

Cook, W. M., Casagrande, D. G., Hope, D., Groffman, P. M. and Collins, S. L. (2004) 'Learning to roll with the punches: Adaptive experimentation in human-dominated systems', *Frontiers in Ecology and the Environment* 2: 467–474.

CoreLabs (2010) CoreLabs website. Online, available at: www.ami-communities.net/wiki/CORELABS (accessed 21 March 2010).

Demeritt, D. (2010) 'Harnessing science and securing societal impacts from publicly funded research: Reflections on UK science policy', *Environment and Planning A* 42: 515–523.

EcoCities (2010) EcoCities website. Online, available at: www.ecocitiesproject.org.uk (accessed 21 March 2010).

Evans, J. (2006) 'Lost in translation? Exploring the interface between local environmental research and policymaking', *Environment and Planning A* 38: 517–531.

Evans, J. (2009) 'The politics of the experimental city', Paper presented at Urban Laboratories: Towards an STS of the Urban Environment Workshop, Maastricht, the Netherlands, November.

Evans, J., Jones, P. and Krueger, R. (2009) 'Organic regeneration and sustainability or can the credit crunch save our cities?', *Local Environment* 14: 683–698.

Evans, R. and Marvin, S. (2006) 'Researching the sustainable city: Three modes of interdisciplinarity', *Environment and Planning A* 38: 1009–1028.

Frissen, V. and van Lieshout, M. (2004) *To User-Centred Innovation Processes: The Role of Living Labs*, Delft: TNO-ICT.

Galanakis, K. (2006) 'Innovation process: Make sense using systems thinking', *Technovation* 26: 1222–1232.

Geels, F. W. (2002) 'Technological transitions as evolutionary reconfiguration processes: A multi-level perspective and a case-study', *Research Policy* 31: 1257–1274.

Geels, F. W. and Schot, J. (2007) 'Typology of socio-technical transition pathways', *Research Policy* 36: 399–417.

Gieryn, T. F. (2006) 'City as truth-spot: Laboratories and field-sites in urban studies', *Social Studies of Science* 36: 5–38.

Gieryn, T. F. (2008) 'Laboratory design for post-Fordist science', *Isis* 99: 796–802.

Grimm, N. B. and Redman, C. (2004) *Central Arizona–Phoenix Long-Term Ecological Research: Phase 2*, Washington, DC: US National Science Foundation.

Grimm, N. B., Faeth, S. H., Golubiewski, N. E., Redman, C. L., Wu, J., Bai, X. and Briggs, J. M. (2008) 'Global change and the ecology of cities', *Science* 319: 756–760.

Hartman, H. (2010) 'Masdar City, Abu Dhabi', *Architects' Journal*, 12 February. Online, available at: www.architectsjournal.co.uk (accessed 19 March 2010).

Henke, C. R. (2000) 'Making a place for science: The field trial', *Social Studies of Science* 30: 483–512.

Henke, C. R. and Gieryn, T. F. (2008) 'Sites of scientific practice: The enduring importance of place', in E. J. Hackett, O. Amsterdamska, M. Lynch and J. Wajcman (eds) *The Handbook of Science and Technology Studies*, 3rd edition, Cambridge, MA: MIT Press.

Hodson, M. and Marvin, S. (2007) 'Understanding the role of the national exemplar in constructing "strategic glurbanization" ', *International Journal of Urban and Regional Research* 31: 303–325.

Hodson, M. and Marvin, S. (2009) 'Cities mediating technological transitions: Understanding visions, intermediation and consequences', *Technology Analysis and Strategic Management* 21: 515–534.

Kidd, C. D., Orr, R., Abowd, G. D., Atkeson, C. G., Essa, I. A., MacIntyre, B., Mynatt, E., Starner, T. E. and Newstetter, W. (1999) 'The Aware Home: a living laboratory for ubiquitous computing research', *Lecture Notes in Computer Science* 1670: 191–198.

Kohler, R. E. (2002) *Landscapes and Labscapes: Exploring the Lab–Field Border in Biology*, Chicago: University of Chicago Press.

Kohler, R. E. (2008) 'Lab history: Reflections', *Isis* 99: 761–768.

Knorr-Cetina, K. (1981) *The Manufacture of Knowledge: An Essay on the Constructivist and Contextual Nature of Science*, Oxford: Pergamon Press.

Knorr-Cetina, K. (1995) 'Laboratory studies', in S. Jasanoff, G. E. Markle, J. C. Petersen and T. Pinch (eds) *Handbook of Science and Technology Studies*, revised edition, Thousand Oaks, CA: Sage.

Kuklick, J. and Kohler, R. E. (1996) 'Introduction', in J. Kuklick and R. E. Kohler (eds) *Science in the Field*, Chicago: University of Chicago Press.

Kusiak, A. (2007) 'Innovation: The living laboratory perspective', *Computer-Aided Design & Applications* 4: 863–876.

Latour, B. (1987) *Science in Action*, Cambridge, MA: Harvard University Press.

Latour, B. (1988) *The Pasteurization of France*, Cambridge, MA: Harvard University Press.

Latour, B. (1998) 'Give me a laboratory and I will raise the world', in M. Biagioli (ed.) *The Science Studies Reader*, New York: Routledge.

Latour, B. and Woolgar, S. (1979) *Laboratory Life: The Construction of Scientific Facts*, Princeton, NJ: Princeton University Press.

Lynch, M. (1985) *Art and Artifact in Laboratory Science: A Study of Shop Work and Shop Talk in a Research Laboratory*, Boston: Routledge & Kegan Paul.

Masdar (2010) Masdar website. Online, available at: www.masdar.ae (accessed 21 March 2010).

Masdar City (2010) Masdar City website. Online, available at: www.masdarcity.ae (accessed 21 March 2010).

MCSP (Manchester City South Partnership) (2008) *Manchester City South Strategic Development Framework*, Manchester: Manchester City South Partnership.

Moore, S. A. and Karvonen, A. (2008) 'Sustainable architecture in context: STS and design thinking', *Science Studies* 21: 29–46.

Niitamo, V.-P., Kulkki, S., Eriksson, M. and Hribernik, K. A. (2006) 'State-of-the-art and good practice in the field of living labs', in K. Thoben, K. S. Pawar, M. Taisch and S. Terzi (eds) *Proceedings of the 12th International Conference on Concurrent Enterprising: Innovative Products and Services through Collaborative Networks*, Nottingham: University of Nottingham.

NSF (National Science Foundation) (2002) 'Long-Term Ecological Research Program twenty-year review', LTERnet Internet. Online, available at: http://intranet.lternet.edu (accessed 10 November 2009).

Oudshoorn, N. and Pinch, T. (eds) (2005) *How Users Matter: The Co-construction of Users and Technology*, Cambridge, MA: MIT Press.

Pearson, J. (2002) *University–Community Design Partners: Innovations in Practice*, New York: Princeton Architectural Press.

Peirce, N. (2010) 'Eyes on Rio as it hosts the World Urban Forum', *Citiwire*, 19 March. Online, available at www.citiwire.net (accessed 21 March 2010).

Perry, B. (2006) 'Science, society and the university: A paradox of values', *Social Epistemology* 20: 201–219.

Pincetl, S. (2010) 'From the sanitary city to the sustainable city: Challenges to institutionalising biogenic (nature's services) infrastructure', *Local Environment* 15: 43–58.

Ravetz, J. (2009) 'Climate change: From global deadlock to local opportunity', *Transforming Management*, 25 November. Online, available at: http://tm.mbs.ac.uk (accessed 21 March 2010).

Schumpeter, J. (1934) *The Theory of Economic Development*, Cambridge, MA: Harvard University Press.

Shields, R. (2008) 'The urban question as cargo cult: Opportunities for new urban pedagogy', *International Journal of Urban and Regional Research* 32: 712–718.

Swyngedouw, E. (2007) 'Impossible sustainability and the post-political condition', in R. Krueger and D. Gibbs (eds) *The Sustainable Development Paradox: Urban Political Economy in the US and Europe*, New York: Guilford Press.

Swyngedouw, E. (2009) 'The antimonies of the postpolitical city: In search of a democratic politics of environmental production', *International Journal of Urban and Regional Research* 33: 601–620.

ULL (Urban Landscape Lab) (2010) Urban Landscape Lab website. Online, available at www.urbanlandscapelab.org (accessed 19 March 2010).

Utterback, J. M. and Abernathy, W. J. (1975) 'A dynamic model of process and product innovation', *Omega* 3: 639–656.

While, A., Jonas, A. E. G. and Gibbs, D. (2009) 'From sustainable development to carbon control: Eco-state restructuring and the politics of urban and regional development', *Transactions of the Institute of British Geographers* 35: 76–93.

10 Municipal bureaucracies and integrated urban transitions to a low carbon future

Alex Aylett

Introduction

Successful responses to climate change will involve broad networks of governance that link multiple actors. But key players in those networks have their own complex internal dynamics. To facilitate low carbon urban transitions we need to understand those dynamics. This chapter focuses on one key player, the municipal bureaucracy, to ask two questions. First, what institutional barriers block integrated municipal responses to climate change? Second, what reforms are necessary to create a municipal bureaucracy able to maintain sustained innovation in the context of changing circumstances? It is based on a case study of how one South African city, Durban, responded to an ongoing national electricity crisis that began in February 2008. This material is drawn from seventy interviews and participant observation carried out both before and during the crisis (between October 2007 and April 2009).

Durban[1] is one of South Africa's leading cities on the issue of climate change. The city's Environmental Management Department (EMD) carries out projects related to food security, carbon mitigation and renewable energy. In partnership with the Tyndall Centre and Golder Consultants, it is conducting advanced downscaling of global climate models – something that is being done in only a handful of other cities in the world. The municipality has also integrated climate considerations into official development planning processes, included them in the performance contracts for high-level management and linked them to broader socio-economic development goals (a link that is itself an important area of research; see Yohe 2001; Pielke 2005; Robinson *et al.* 2006).

While beginning this process of integration, Durban has been the site of a triple crisis of energy shortages, food price hikes and extreme weather. These overlapping stresses make it an ideal window into what influences the ways in which complex socio-political and technological systems respond to crisis, and how those responses can be improved.

The perfect storm

In early 2008, a major energy crisis hit South Africa. Generation capacity had stagnated for over a decade as a result of a half-hearted attempt to privatize electricity generation. Simultaneously, post-apartheid social and economic development objectives, focused on economic growth and increased domestic electrification, drastically increased consumption. Until recently, electricity tariffs in South Africa were half the rate charged in Canada, its next closest competitor in the global electricity market.

Growing consumption ate into reserve generation capacity to the point where plant maintenance was next to impossible. There was no room in the system to compensate for the loss of electricity that happens when a plant is brought off-line for inspection and repairs. In January 2008, this unsustainable situation collapsed when plant failures caused uncontrolled blackouts that swept across the country. In the month that followed, a system of managed rolling outages (load shedding) was put in place. Thousands of jobs were lost and job creation stagnated – this in a country with an unemployment rate of 26 per cent (Stewart 2008). The total losses to the South African economy are estimated at 50 billion rand (US$6.07 billion) (Patel 2008).

This occurred in the context of a global food price crisis and rapid increases in the cost of other fuels. In South Africa, the cost of staple foods such as bread and cornmeal rose by 25 per cent during the first six months of 2008 (Hurd 2008), the cost of a tank of gas rose by close to 16 per cent (IOL 2008) and inflation hit 11 per cent (the first time for five years it had risen into double digits) (ibid.).

The local context in Durban was even rougher. One year earlier, a major coastal storm severely damaged both municipal infrastructure and private property. Damages to public property in Durban alone were estimated at 84 million rand (US$11.5 million). In addition, record-breaking rainstorms hit the city in 2007 and 2008. Four people were killed in the flooding, one thousand displaced, and roads, bridges and other infrastructure damaged (News24, 2007, 2008).

Local NGOs and city officials have begun linking many of these events directly to climate change. These links are still uncertain. But these events have created in Durban the perfect storm for studying a municipality's ability to respond to the multiple stresses that a changing climate is likely to put on coastal municipalities. To better understand the factors that influence organizational responses to these kinds of stresses, this chapter first explores the linked analytical concepts of 'trained incapacity' and 'organizational culture'. It then compares the organizational cultures of two of Durban's municipal departments for a more concrete look at the institutional factors that can foster or fetter innovative responses to unstable circumstances.

'If you *are* a hammer . . .': trained incapacity and organizational culture

Over the course of interviews with municipal leaders, department heads and civil society groups, it became clear that even though we talk about 'the city', this vision of a unified purposeful actor makes little sense. 'The city' is made up of numerous entities, each one constrained and motivated in its own ways. Engaging with this more uneven reality of how power and purpose are distributed is essential if we want to understand either Durban's response to the energy crisis or how responses to climate change are now playing out, or will soon play out, in cities more generally.

Understanding change within these types of complex systems is the core concern of both socio-technical system theory (STS) (Berkhout 2002; Geels 2004; Smith *et al.* 2005) and new institutionalism (NI) (March and Olsen 1989; Peters 2005; Young 2002). In fact, there is increasing common ground between the two approaches as STS has begun to critique rationalistic simplifications of power and agency within large organizations. Cities, as complex bureaucratic entities that are also responsible for large technological networks, are particularly well suited to an analytical framework that draws from both NI and STS.

I do not wish to gloss over the important differences between STS and NI. With its analytical focus on how to create change in large 'landscapes' of socio-technical tools and practices, STS strives to understand and perfect the ways small-scale change in protected niches can bring about large-scale shifts. NI is interested in systemic change, but it tends to be descriptive rather than prescriptive. Its unit of analysis is also more closely delimited to the boundaries of individual organizations and their institutional histories. These differences make them quite complementary, with NI providing a necessary look inside the institutional dynamics of the 'social' components of socio-technical systems.

There are two key points of agreement between these bodies of work that are relevant to the case study that follows. The first is that individual values, power and agency are deeply influenced by organizational structures (this in contrast to rational-actor models, which commonly assume that these variables are formed independently of institutional context). This is a point extensively developed by NI but also well discussed in STS (see Smith *et al.* 2005). The second area of agreement is that even apparently stable systems are in fact constantly undergoing incremental changes, as established practices make sense of and respond to changing conditions. In STS, discussions of change emphasize the search for efficiencies within established systems. NI's more political focus is on how actors at various levels strive to protect established socio-political networks, power distributions and systems of meaning in the face of changed circumstances.

These organizational delimitations of individual and collective agency and creativity are of course linked. The stability of the existing regime rests on its ability to shape individual thoughts and actions in a way that makes them both coherent and predictable across the organization as a whole. But what does that imply for the process of attempting to create change that pushes past the limits of existing institutional structures? To answer that question, I would like to focus on two concepts whose genealogy is part of the intellectual heritage that has influenced both STS and NI. The terms 'trained incapacity' and 'organizational culture' have their roots in the late-nineteenth-century work of economist and sociologist Thorstein Veblen and have grown into a conversation that winds its way through John Dewey, Kenneth Burke (1935), Robert K. Merton (1940) and, more recently, the work of Erica Schoenberger (1997).

Trained incapacity

At its most basic, trained incapacity refers to a simple argument: through our training, and subsequent professionalization, we are schooled to use specific tools and analytical systems to define and accomplish our goals. These ways of seeing and acting on the world around us can also produce blind spots that incapacitate us when solutions, or even problems themselves, fall outside of these professional limits (Veblen 1898, 1914, 1918). Specialized training produced what Veblen called a 'widening . . . field of ignorance' (1918: 152). The more expert one becomes, the less able one is to respond to, or even perceive, issues that fall outside one's area of expertise.

At a more general level, Veblen held that some form of trained incapacity was innate to the human condition. He argued that the human species is by nature industrious and driven to define and pursue goals. In the pursuit of these goals, we develop habits (reinforced, according to Veblen, by processes of natural selection) to guide our actions and increase our efficiency. These habits in turn give rise to patterns of thought and systems of value based on what we are most accustomed to seeing and doing:

> What men [*sic*] can do easily is what they do habitually, and this decides what they can think and know easily. They feel at home in the range of ideas which is familiar through their everyday line of action. . . . What is apprehended with facility and is consistent with the process of life and knowledge is thereby apprehended as right and good.
>
> (Veblen 1898: 195)

Habitual material practices create and embed the individual within larger ways of knowing the world (epistemologies) and ways of judging it (values and ethics). While these may guide action in certain circumstances, they frequently act as blinders, hiding other realities and courses of action from view when conditions change.

Organizational culture

Burke and then subsequently Merton pick up this concept, broadening the focus of Veblen's work to look at how the propensity towards trained incapacity at the individual level can be shaped – even intentionally fostered – by the larger organizational culture of the institutions that surround them. Merton (1940), who focuses specifically on bureaucratic structures like those that we still find in the contemporary city, makes clear both the institutional causes and the costs of trained incapacity. He argues that to operate effectively, a bureaucracy must achieve a high degree of consistency and social conformity within its ranks. This is made possible only by instilling 'appropriate attitudes and sentiments' (ibid.: 562) among those who work within the organization. An organizational culture of shared systems, rules and procedures is put in place to ensure that material practices, epistemologies and ethics are held in common across the organization.

Organizational cultures do not necessarily have to encourage trained incapacity. But often they do, given that it can be a useful tool in the process of creating uniformity and predictability within a large bureaucracy. The negative outcome of this is that rules, initially put in place as a means to an end, are seen as ends in and of themselves. This creates an almost ritualistic adherence to rules for their own sake. This is the stuff from which 'bureaucratic nightmares' are made. The weight of these systems makes adjusting quickly to changing conditions all but impossible. This can be the case even when the new circumstances mean that old ways of doing business jeopardize not only the aims of the organization but its very existence.

In Schoenberger's (1997) work on US automotive companies, we see this problem of trained incapacity taken almost to the point of corporate suicide. She provides several fascinating case studies of the organizational cultures that prevented American firms from responding to a changing market and new Asian competitors. Beyond this, she makes a valuable contribution by questioning the determinism implicit in her predecessors' arguments. She argues that trained incapacity, and the type of organizational culture that encourages it, are dynamic, not static. We need to understand not simply why organizations resist change, but also how and why particular changes are accepted while others are resisted. Implicit in this dynamism is the opportunity for positive adaptive change.

The problem, she argues, is not so much that information about a changing world is hard to come by. Rather, echoing arguments made by Haraway (1991), corporate knowledge is always situated knowledge:

> This situatedness may produce what could be thought of as a predisposition to systematically read evidence about the state of the world in certain ways – or, alternatively, to systematically misread certain kinds of evidence. Yet again, it may imply systematically ignoring certain kinds of evidence altogether so that information, though available, become uninformative.
>
> (Schoenberger 1997: 125–126)

This does not mean that threats are not perceived, but that they are interpreted in such a way as to fit within the already established structures in place within a given organization. Managers and employees will choose responses that preserve the value of specific material infrastructure, as well as the social and cultural assets manifest in specialized skills, administrative procedures and protocols. This instinct to defend existing socio-technical assets, and the structures and relationships that they hold together, has a strong influence over the way responses to new conditions are selected and implemented.

This is particularly the case for those who are most powerful within a current structure. The position of senior managers, for example, depends precisely on the current arrangement of socio-technical assets (which they themselves have put in place), as well as the broader social and political networks that these arrangements hold in place. In addition, they are the ones who hold a monopoly on the power to creatively imagine both the strategic vision of the company and its goals, as well as to define the vision of the world within which they operate. Hence, any drastic change is unlikely – even if staying the course will have increasingly negative effects on the organization as a whole.

But this is not inevitable. Rather than harnessing bureaucratic structures to coordinate an intentional expansion of the human propensity for trained incapacity, we can use them to open up possibilities rather than close them down. Organizational cultures can be created that encourage innovation. Rather than maintaining the monopoly on imaginative action among senior management (those in many ways least likely to exercise it fully because they are so deeply embedded in the current state of affairs), Schoenberger argues that the power to act creatively needs to be distributed more broadly throughout the organization. Given the situatedness of all knowledge, and the limits that trained incapacity places on all of us, effective responses to radically changing conditions can only come when strategic vision is produced through a negotiation between multiple actors 'whose social and cultural locations and mix of social and material assets are different' (Schoenberger 1997: 229). For her, this extends to a much more radical project: 'This implies a profound reordering of power relations within the firm and between the firm and other social actors (communities, unions, other firms, etc.)' (ibid.).

As we will see in the case studies that follow, this type of distribution of empowered creativity with an organization can have powerful results. Certain departments within Durban's municipal bureaucracy have established a profoundly different organizational culture. They have put in place a culture that distributes empowered creativity much more broadly within the organization, and embraces the new social and material practices that are co-constitutive with innovation and the ability to respond to changing circumstances. But in other cases, established institutional patterns can be very resistant to change.

The electricity crisis: the initial response

Even though it was widely known that the national public electricity utility (known as ESKOM) had allowed its electricity supply to become severely overstretched, when the crisis hit it took everyone by surprise. For eThekwini Electricity (EE), the municipal electricity distributor, load shedding forced it to rapidly reshape its role within the system. Traditionally, EE acted as a bulk purchaser of power from the national grid, and a local distributor and sales agent to clients within the municipality. It was an intermediary between ESKOM and local clients, and it also managed and maintained the local grid. Prior to the crisis, its priorities were to maximize electricity sales by actively competing to displace the use of coal and gas among potential clients (industrial and residential). All of this was linked to larger national development goals of increasing economic activity and providing access to basic services to all South Africans.

With the crisis, it found itself suddenly asked to do the opposite: instead of supplying more demand for ESKOM, it had to supply less. In the early days of the crisis, the municipal utility would receive word from its national counterpart that a certain number of megawatts needed to be shed from the city's power consumption. It then had five to ten minutes to use the switching system built into the local grid to select and shut down entire portions of the city. Technicians had to meet the needs of the national grid while doing the least damage to the needs of local residents and businesses and not unfairly closing off power to some sections of the city more than others. It was a tricky balance of precise technical actions and a complex socio-economic and racially charged context of energy politics within the city. After some delay, these rolling blackouts were orchestrated and load-shedding schedules were published in local papers.

In interviews one month after the onset of the crisis, the main concern of the key spokesperson within EE was with the need to increase the reach and complexity of its switching system. Deena Govender, Manager of Commercial Engineering and Marketing, is effectively second in charge within the utility. He was tasked with developing the response to the energy crisis. In interview, he explained how the existing network of switches was spread unevenly across the municipal area. It offered at best neighbourhood-by-neighbourhood control. Hospitals and other key service points within the city were often impossible to untangle from larger non-essential residential loads within the city.

EE's main preoccupation at the time was to solve this by extending centralized switching control right down to the level of individual household appliances. Other cities have used this smart switching technology, in conjunction with time-of-use metering (which the city considered later on), as components in a localized smart grid able to synchronize and create incentives for the introduction of local renewables into the grid. In Durban, the two technologies were completely divorced from any role they might play in changing the way the city procured its electricity.

By installing smart switching technology, already on the market, EE planned to have direct control over large non-essential household loads (air conditioners, water heaters and pool pumps). The cost of the programme, both financial and in terms of drain on management capacity, would have been considerable. But it was accepted as a clear extension of the municipal distributor's current practices. When EE was confronted with a clear need for change, a course of action was found that preserved the value of the organization's current professional capacity and replicated established models for built infrastructure (March and Olsen 1989). The inertia of trained incapacity allowed the distributor to extend a well-trodden path into unfamiliar territory. It also cut off any serious discussion of other options.

The potential for local renewable energy generation, for example, was almost completely overlooked. Govender and others within the organization were quick to dismiss any support for renewables from their side. They were explicit about the fact that they were not 'in the business' of generating power, that it was 'not part of their core responsibilities' and that (supposedly) unresolved technical issues made decentralized renewables impossible in the local context. They pointed to concerns over the quality and safety of the power that renewables would put into the grid.[2] In reality, both of these are red herrings; reliable mechanisms to regulate both are readily available on the market – some manufactured by South African companies.

This blockage was not caused by a lack of information. In fact, the city's Environmental Management Department (EMD), in collaboration with a consulting firm, had produced a municipal energy strategy. The document outlined the potential for renewable energy within the city, and linked the development of local green energy to a proactive approach to climate change. A greater presence of local renewables would mitigate the city's emissions and increase the resilience of the city's energy supply to climate-related disruptions of the national grid. In discussion with EE, however, Jessica Rich (then managing the EMD's energy programme) found that the same refrain emerged: ' "You don't understand," they would tell me, "this [renewables] just isn't what we do!" ' (Rich, interview 30 March 2009). As Schoenberger argued, it is not that information was scarce. It was that established habits of thought and action made it impossible to accept available information and transform it into a new vision or strategy for the organization.

One year later, the global economic crisis had significantly reduced industrial demand for electricity. Load shedding had stopped. Fundamentally, though, the nation's electricity supply remained the same. To prevent future blackouts once the economy recovered, EE was under pressure to continue to produce residential energy savings. Its energy efficiency drive had grown to include the beginnings of a solar hot water programme and a programme to replace high-pressure sodium street lights with LED lights. Many things had changed, but the core focus on energy efficiency and a fine-grained switching system remained (although it had yet to be implemented).

EE's position on renewables had also stayed the same. Concerns over safety and quality remained – although both apparently had gone uninvestigated. And an insistence on the limits of its 'core mandate' persisted:

> How much can we do? Our core business function is to buy electricity and to distribute it as effectively as we possibly can. We are not generators of electricity. Our core business is not to generate electricity; it is to distribute and to transmit electricity. From that perspective, how much can we contribute to reducing our carbon footprint?
>
> (Anonymous eThekwini Electricity official, interview 2009)

While not uncommon, this strict adherence to official job descriptions is not unavoidable. As we will see, important changes often come from actors who see their stated mandate as the starting point and not the limit of their work. What it points to here is the need to look beyond the technical solutions proposed by the EE, to the organizational culture that selected some solutions while excluding or ignoring others. The emphasis on switching and efficiency is a strong example of the way in which an organization that fosters trained incapacity responds to shocks – even, in this case, radical ones – by repurposing and adapting established tools and practices.

Switching systems and the skills and response capacities linked to them have always been central to the department. Similarly, efficiency has a long lineage in the department. As a former sales officer explained, electrical efficiency was one of the tools it used to encourage large industrial customers to switch from coal or gas (Sing, interview 24 March 2009). Working with clients to improve their efficiency would reduce EE's costs, making it more likely that they would electrify more of their operations. The repurposing of a tool originally used to increase demand therefore does not raise any real challenges to established social or material practices.

Further, EE's role as an intermediary in the sale of electricity to local customers is solidified through powerful professional (and no doubt on an individual level, personal) relationships with the national energy supplier, ESKOM, and the energy generation model that ESKOM represents. There is the perception that local engagement with renewables must wait for national bodies to take the initiative: 'It is not driven nationally. Municipalities do not have resources to go and investigate these things. We are here to provide electricity. That is our full-time job' (Govender, interview 13 March 2009). Officials repeatedly emphasized a particular hierarchy of generation, bulk purchase, local distribution and final consumption that to them was an immovable aspect of how they did business:

> In terms of generation we as a municipality do not have direct control. . . . Our core business function there again is to distribute and transmit as effectively as we possibly can. You guys talk about mitigation, but how much can we do within our core business function to support mitigation? The real mitigation stuff has to happen at the top [ESKOM].
>
> (Anonymous eThekwini Electricity official, interview 2009)

EE, as a department, is deeply embedded in social and technical networks that involve both built infrastructure and relationships with other powerful players. The inertia built into these social and technical networks complicates change and innovation. But it would be wrong to paint EE as completely without options. As we have seen, it has the ability both to modify physical infrastructure and to tailor energy tariffs to influence patterns of consumption and encourage local generation. These are two key levers through which municipalities can facilitate and create incentives for decentralized renewable energy.

That this potential has stayed dormant brings us back to a point made by Veblen in the late 1890s about the links between actions, knowledge and values. As he argued, habits of action (ways, for example, of managing a complex power grid) embed practitioners in ways of knowing and understanding the world, and from this understanding comes a sense of values or, in his words, of what is 'right and good' (Veblen 1898: 195). This question of values was a strong current in my interviews with EE employees. There seemed to be an underlying sense that it was simply not appropriate, not 'right', for them to be involved with promoting the generation of power at the local level.

The narrow interpretation of EE's mandate is further reflected in the lack of connections it has made between energy projects and other local development objectives. The thousands of installation and maintenance jobs created by the adoption of technologies such as solar hot water, for example, could be of enormous benefit to the local economy. Decentralized energy projects can also facilitate access and improve the quality of service in difficult-to-service areas. They simultaneously answer linked environmental, energy and quality-of-life related development goals. Although discussed by other municipal employees, none of these potential synergies came up in interviews with electricity officials.

Faced with a sudden crisis, the department responded as many of us would: its staff used the tools familiar to them and attempted to extend and modify past procedures to new conditions. This is, as the work of Veblen, Merton and Schoenberger helps to make clear, a predictable and in some ways very human response. But it leaves key opportunities unrealized and will not generate the types of changes that Durban needs to resolve its energy problems, decrease its emissions or become more resilient. We need to look for ways in which this institutional and personal inertia can be overcome.

Beyond business as usual

The EMD, which produced the city's municipal energy strategy, was willing to respond creatively to energy and climate issues up to a point, but refused to become involved with implementation. Its staff see their role as providing information to catalyse action in other departments. In fact, the reports and discussions they facilitated around alternative energy paths for the city were crucial in getting senior-level support for the creation of an independent Energy Office, launched in February 2009. While it is an encouraging step, the new office is hampered by a lack of strong leadership and a difficulty in attracting qualified staff.

For the foreseeable future, it will only be able to act as a coordinating body for projects already being undertaken by other agencies. But here we find ourselves in the same bind as before. Given that no other departments have the mandate to engage with renewable energy projects, the new Energy Office will find few partners – unless of course some agencies are willing to go beyond what is formally contained in their job descriptions.

There are in fact a variety of alternative energy projects being run by departments within the city. These range from in-pipe hydro and micro-hydro in the city's water reticulation system, to biodiesel from algae projects to methane gas capture at two of the city's landfill gas sites (one of which is Africa's largest UN Clean Development Mechanism project). All but the latter are being run by the water and sanitation department (eThekwini Water Services, EWS).

In terms of actual energy yield, their flagship programmes are a complementary pairing of biogas and hydropower projects that could criss-cross whole sections of the city. The city's hilly terrain generates excess pressure within the fresh-water distribution system. Water from dams descends into the city, building pressure that needs to be cut by as much as two-thirds in some parts of the system. It is currently dissipated as noise and heat by mechanical pressure release valves. To make better use of it, hydro and micro-hydro generation turbines are going to be integrated directly into the piping system. In total, they will yield enough power for between 10,000 and 30,000 low-cost houses (7–22 MW), depending on the extent of the roll-out.

EWS also has plans to upgrade and expand the existing generation systems that capture gas produced by the biodigesters that break down sewage. While continuing to meet the needs of the treatment works themselves, the plants will store biogas to generate electricity, which can then be sold to the grid at peak times. In partnership with AGAMA energy (a South African renewable energy consulting firm), they are putting in place smaller-scale applications of the same system in peri-urban communities. These setups make it possible to provide low-cost water-borne sewerage, to treat effluent on site, to produce gas that can be used for cooking and to generate a high-grade fertilizer that can be used for local agriculture. This reduces the city's environmental footprint, increases its energy security and boosts the economies of its poorer communities.

These projects are only two out of a long list, including plans to create biodiesel from algae grown on sewage. While they are interesting, more interesting is how a department with no mandate to generate energy came to see these opportunities and act on them. While many of the officials that I talked to in the city were generally interested in 'doing something about climate change', many – not only those within EE – explained that it simply wasn't part of their job description.

In EWS, the answers were different. Employees and senior management perceived their core mandate as defining where their work began, not where it ended. Your job description was 'the boring part of your job', and from management on down, there was an organizational culture that rewarded innovation and risk taking. Speedy Moodliar, EWS's Manager of Planning, told me, laughing:

> I can design a water main with my eyes closed. There is nothing more anyone can teach me about infrastructure planning in terms of water or designing for water. I can do it. In life you need more than that, you need something that is going to interest you and stop you from yawning.
>
> (Moodliar, interview 26 March 2009)

So how can we account for this difference? There are a few relevant contextual facts. Unlike EE, EWS is not tied into business relationships with other powerful agencies. There is no equivalent to ESKOM within the field of water and sanitation, and hence changes to EWS operations do not affect anyone else's business plan. Second, EWS is not caught in the type of national-level crisis that has hit electricity. Instead of responding reactively and under extreme pressure, EWS has the room to think and plan more calmly and creatively. Finally, EWS has raw materials (water pressure and effluent) that can be used as the basis for energy generation.

But beyond these differences, there is something else at work. In interview with Neil Macleod, the head of EWS, it became clear that the organization was managed very differently from other municipal departments. Everything seemed to be being done to encourage innovation and to fight the overspecialization that creates trained incapacity. At an individual level, employees are pushed to see their job descriptions as the minimum level for their work. Beyond that, the organizational culture of the department encourages employees to think critically about how to achieve their objectives and how to address problems:

> You are expected to take responsibility and to challenge everything that you do about your job. If you see an opportunity for innovation but can't initiate the solution yourself, then come and talk to me and let's find a way to break down the walls out there that are stopping you from achieving it. . . . There's something else: I always tell people, 'Don't come with a problem, come with a solution. Here is the issue, here is how we can solve it. What do you think? OK, try it out.' Always going to your boss for a solution becomes very limiting both for you and for me and the other top management.
>
> (Macleod, interview 27 March 2009)

At the level of the organization as a whole, similar principles apply. Cross-level meetings bring staff together to share challenges and collectively brainstorm solutions: 'We get everyone from clerical right up to senior management in the same room looking at the same problem and bringing their own perspectives in' (ibid.). EWS also holds a monthly sustainability lecture series, with three or four invited speakers covering a very wide range of sustainability-related topics.[3] Space for discussion at the forums is open and informal, and brainstorming here was the starting point for EWS's biodiesel projects.

Taken together, these initiatives have created a very particular structure within EWS. The standard organizational structure of a municipal department approximates that of a hierarchy of siloized sub-units (see Figure 10.1). In this type of organization, senior management determine the direction and mandate of the organization as a whole. Responsibility for fulfilling parts of that mandate cascades down through different units, divisions and individuals, each with their own sub-area of action and influence. Communication flows primarily (but not exclusively) from top to bottom, and employees have little chance to communicate between units or to influence the direction of the organization as a whole. Finally, the boundaries of the organization are relatively solid, offering little chance for collaboration with outside organizations or departments.

Distancing itself from this siloized and closed model, EWS created an internal structure that is both more integrated and more open (see Figure 10.2). Although the structure is still hierarchical, inclusive intra-departmental forums encourage communication and exchange between sub-units. Encouragement to form partnerships and think creatively means that individual influence extends beyond the confines of official divisions and can spread out, potentially contributing to the mandate of the organization as a whole. This makes it possible to identify synergies and shared interests with other actors both inside and outside the organization. A more permeable organizational border brings in outside input and

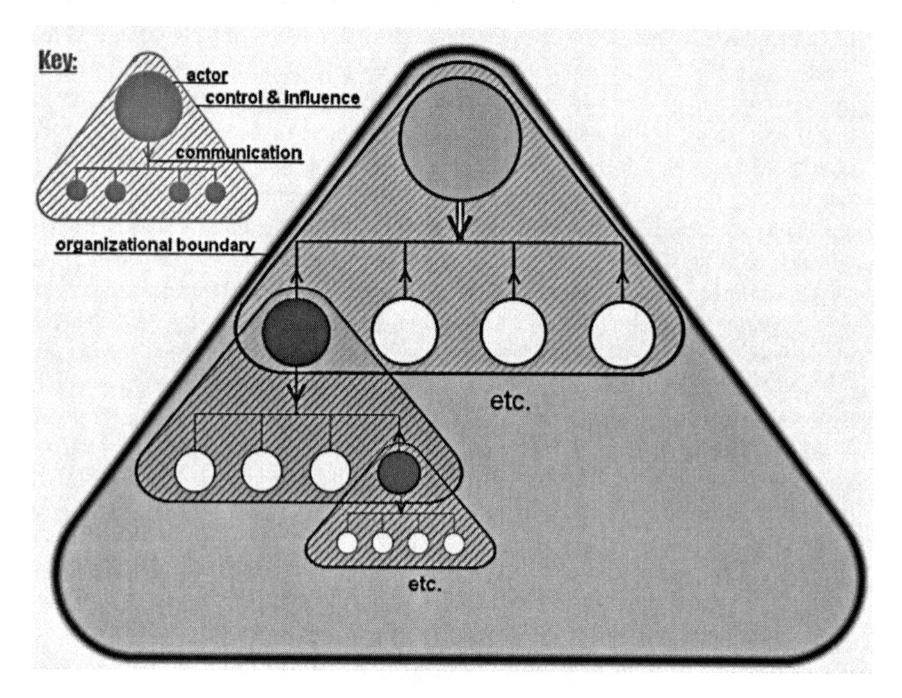

Figure 10.1 A hierarchical, siloized and closed-loop organization.

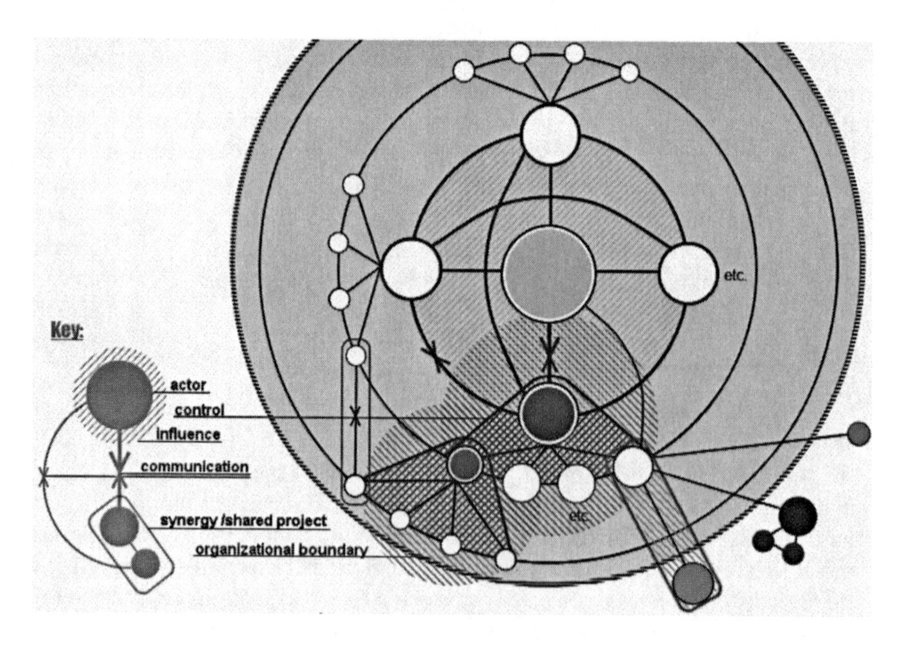

Figure 10.2 A hierarchical, integrated and open organization.

information to allow for the creation of a strategic vision that incorporates input from a variety of situated knowledges. These exchanges can also develop into partnerships around specific projects that bring increased resources and expertise to the organization while meeting a variety of developmental goals.

The inclusive redistribution of empowered creativity encouraged by EWS is a significant shift from the way in which municipal departments are traditionally run. Its practices embody a more modest version of Schoenberger's call, discussed above, for the profound reordering of power relations within organizations, and between them and other social actors.

Officially, senior managers like Macleod justify all of this by pointing to their bottom line: 'All our energy from waste projects, they are purely to try and minimize our electricity bills. The fact that it benefits climate change is incidental to us' (Macleod, interview 27 March 2009). As well as emphasizing synergies between these projects and EWS's core objectives, Macleod and Moodliar both stressed the opportunities that exist to help meet broader municipal objectives in areas of economic development, improved health and quality of life. But just below the surface, it is clear that senior managers like Macleod and Moodliar are motivated by the rewards of innovation itself. Rather than use departmental structures to encourage specialization and trained incapacity, they have engineered it to do the reverse.

Macleod speaks explicitly about the need to create a culture of innovation within the department. There are still strong pressures on employees to conform to certain norms, but innovation has been built into those norms: 'You tend to find people that like change and innovation and that kind of life stay [in EWS]. . . . There are lots of people who don't and they either leave or are asked to leave' (Macleod, interview 27 March 2009). Echoing Schoenberger's arguments covered earlier in the chapter, the strength of the department comes from the fact that it empowers individual creativity within the organization, rewards risk taking and actively seeks out diverse opinions and viewpoints to feed ideas into its projects. Instead of a creating recursive loops between skills and practice, EWS is a learning organization capable of taking on board new ideas that in turn guide new material practices.

The value of having a 'champion' for sustainability initiatives is often discussed. What this case study shows more clearly is that more than simply driving change individually, a champion in a position of control over institutional structures can create a culture that encourages multiple actors to seek out and develop alternative courses of action.

Conclusions

Trained incapacity and a conservative organizational culture within Durban's electricity distributor, coupled with the pressures of a massive electricity crisis, were drivers for institutional innovation within the municipality – but not where one might have expected it. The municipal electricity utility refused to engage with any facet of the design, implementation or creation of incentives for renewable energy. The fact that these issues are currently being dealt with by the newly formed Energy Office, even if it is only in its earliest stages, shows the value of having an independent institutional home for issues such as energy reform and climate change. Climate change-related policies are challenging precisely because they cut across the jurisdictions of multiple municipal institutions – all already embedded in their own institutional cultures, epistemological structures and material practices.

But a dedicated climate change, energy or sustainability office will need to find willing partners within the city. It needs to be joined by departments, or people within departments, able to drive innovation and change on their own. EWS's example shows one way in which organizational culture can be harnessed to loosen the constraints imposed by trained incapacity and empower creativity.

As mentioned briefly in the introduction to the chapter, via its development planning process the municipality has already begun the process of trying to link development objectives and encourage interdepartmental collaboration. At least on paper, responding to climate change occupies a key place within those efforts.

The contrast between the organizational cultures in EE and EWS shows some of the challenges these efforts will face in practice. Increasing a department's ability to respond to changing circumstances involves a far-reaching shift in the conventions and structures that traditionally govern large organizations. It implies a willingness to break senior management's monopoly on creative imagination, to foster risk taking and a diversity of opinions with the organization, and to actively seek outside input from partners, whether they be communities, consultants, universities or international agencies. These new relationships will bring with them their own costs (in terms of time, if nothing else), but they will also help rejuvenate the intellectual capital within the organization and break patterns of mental and material path dependency.

If we knew precisely what climate change had in store, one could argue that what we need is a transition that adapts the practices of existing organizations to new conditions. The way that power, creativity and agency are distributed could remain the same, once the initial reorganization had taken place. But rather than a shift from one relatively steady state to another, climate change holds the promise of a prolonged period of instability and change. To face that, we need to foster organizations that are able to maintain sustained periods of learning and creative action. That type of adaptive and resilient response goes beyond any single 'shift'. It means moving away from organizations that instrumentalize trained incapacity to ones that empower creative action and collaboration to meet shared goals. And more than just formal modifications to the institutional structures and built infrastructure of our cities, we need to put in place new organizational cultures that will make the rapid transitions and sustained adaptive responses possible.

Acknowledgements

I would like to thank the Trudeau Foundation, the Social Sciences and Humanities Research Council, and the International Centre for Sustainable Cities for their support.

Notes

1 Following a merger with its extended metropolitan area, Durban became officially known as the municipality of eThekwini. 'Durban', however, is still the name most commonly used to refer to the city, while eThekwini is used for official municipal documents and departments. This article follows those conventions.
2 When you feed electricity back into the grid, you have to ensure both that it is cycling at the right speed and that it shuts off if the grid goes down. This is to make sure the electricity does not disrupt the grid, and that it does not electrocute linemen going out to fix technical problems.
3 The forum I attended had presentations on seed swapping and food security in India, green building design principles, and the sanitary composting of sewage material to create agricultural fertilizer.

References

Berkhout, F. (2002) 'Technological regimes, path dependency and the environment', *Global Environmental Change* 12: 1–4.
Burke, K. ([1935] 1984) *Permanence and Change: An Anatomy of Purpose*, 3rd edition, Berkeley: University of California Press.
Geels, F. W. (2004) 'From sectoral systems of innovation to socio-technical systems: Insights about dynamics and change from sociology and institutional theory', *Research Policy* 33: 897–920.
Haraway, D. (1991) *Simians, Cyborgs, and Women: The Reinvention of Nature*, London: Free Association Books.
Hurd, E. (2008) 'Kids dumped because of food hikes', Sky News online: http://news.sky.com/skynews/Home/World-News/Food-Price-Increases-Blamed-For-Rise-In-Abandoned-Children-In-South-Africa/Article/200807415062043 (accessed 1 August 2008).
IOL Staff (2008) 'Petrol creeps to R10 a litre', *Tribune* online: www.iol.co.za/index.php?set_id=1&click_id=13&art_id=vn20080224091320103C782033&page_number=1 (accessed 24 February 2008).
March, J. G. and Olsen, J. P. (1989) *Rediscovering Institutions: The Organizational Basis of Politics*, New York: The Free Press.
Merton, R. K. (1940) 'Bureaucratic structure and personality, *Social Forces* 18 (4): 560–568. Online, available at: www.jstor.org/stable/2570634 (accessed 1 June 2009).
News24 staff (2007) 'KZN storm costs mounting', News24 online: www.fin24.com/Business/KZN-storm-costs-mounting-20070329 (accessed 30 March 2007).
News24 staff (2008) 'KZN floods break records', News24 online: www.news24.com/SouthAfrica/News/KZN-floods-break-records-20080619 (accessed 19 June 2008).
Patel, S. (2008) 'Whistling in the dark: Inside South Africa's power crisis', *Power* online: www.powermag.com/business/Whistling-in-the-dark-Inside-South-Africas-power-crisis_1488.html (accessed 1 November 2008).
Peters, B. G. (2005) *Institutional Theory in Political Science: The New Institutionalism*, New York: Continuum.
Pielke, A. R., Jr (2005) 'Misdefining "climate change": Consequences for science and action', *Environmental Science & Policy* 8: 548–561.
Robinson, J., Bradley, M., Busby, P., Connor, D., Sampson, B. and Soper, W. (2006) 'Climate change and sustainable development: Realizing the opportunity', *Ambio* 35 (1): 2–8.
Schoenberger, E. (1997) *The Cultural Crisis of the Firm*, Oxford: Blackwell.

Smith, A., Stirling, A. and Berkhout, F. (2005) 'The governance of sustainable socio-technical transitions', *Research Policy* 34: 1491–1510.

Stewart, T. (2008) 'Economic meltdown in South Africa', *Sunday Times* online, 7 February 2008. Online, available at: www.ever-fasternews.com/index.php?php_action=read_article&article_id=738 (accessed 10 February 2008).

Veblen, T. (1898) 'The instinct of workmanship and the irksomeness of labor', *American Journal of Sociology* 5: 188–201.

Veblen, T. (1914) *The Instinct of Workmanship and the State of the Industrial Arts*, New York: Macmillan.

Veblen, T. ([1918] 1957) *The Higher Learning in America*, New York: Sagamore Press.

Yohe, G. W. (2001) 'Mitigative capacity: The mirror image of adaptive capacity on the emissions side', *Climatic Change* 49: 247–262.

Young, O. R. (2002) *The Institutional Dimensions of Environmental Change: Fit, Interplay, and Scale*, Cambridge, MA: MIT Press.

11 Community-led urban transitions and resilience

Performing Transition Towns in a city

Amanda Smith

Introduction

In the United Kingdom, we have witnessed numerous efforts, at a range of governance levels, to facilitate transitions towards low carbon economies and places. For instance, the Climate Change Bill of 2008 is underpinned by a key aim of facilitating the transition to a low carbon economy. Increasingly, the claims for these transitions have been grounded in concerns for managing risk, ecological and energy security, fostering resilience and adapting to changing (global and local) environments. With more than half of the world's population now living in urban areas, and cities accounting for 80 per cent of greenhouse gas emissions (Hodson and Marvin 2009), cities are often seen as crucial sites for responding to the dual issues of climate change and peak oil, while also becoming more sustainable in nature (Betsill and Bulkeley 2007; Moser and Dilling 2007; Giradet 2008). A number of towns and cities in the United Kingdom (and across the globe) have taken initiatives to move the agenda further, such as the Nottingham Declaration on Climate Change[1] and the ICLEI's Cities for Climate for Climate Protection Campaign.[2]

However, top-down, centrally led initiatives have seen limited progress in terms of galvanising public support or bringing about local level changes in attitudes and behaviours (Barr 2008). The issues associated with climate change, air pollution and resource crises are complex and intertwined with interactions between society and the environment and technology (Bickerstaff and Walker 2003), and hence top-down approaches may have limited impact in communities that have a lack of faith in institutions of the state, a weak sense of personal agency and little trust in information provided by the state (Macnaghten and Urry 1998). Conversely, the grassroots Transitions Town social movement, initiated in 2006 in Totnes, Devon, by the permaculturalist Rob Hopkins, has seen considerable uptake within the United Kingdom and across the globe. The movement (or culture, as it is sometimes described) focuses on supporting 'community-led responses to peak oil and climate change, building resilience and happiness' (Hopkins and Lipman 2009: 7). To date, there are 277 Transition Initiatives (they range in scale including islands, villages, towns and cities) in countries across the world, including Australia, Japan, Germany, Ireland, New Zealand, Chile, the Netherlands, Canada and the United States.

The speed with which the movement has spread and its ability to speak to a variety of geographical scales and often divergent communities of interest provides useful material for addressing understandings of transitions, the ways in which cities might shape transitions and methodologies for researching urban transitions. The aim of this chapter is to employ a discursive analytical approach to explore the performative nature of the discourses employed by the Transition Towns movement. This reveals why and how the movement has been adopted in the case study under analysis but also reveals the problematics of attempting to rescale conventional Transition Town discourses at an urban level. Indeed, the chapter focuses on discursive forms of power and specific forms of social capital involved in such grassroots movements and their scalar politics (Putnam 2004; Demeritt 1998). This allows us to consider the transformative potential and obduracy of transitions in urban contexts.

This chapter is divided into four sections. The first section outlines the methods and theoretical underpinnings of the research approach; the second establishes the history of the Transition movement and provides some context for the performative discourses, before the subsequent section explores the key discourse of resilience. In the fourth section, the chapter turns to examine emergent themes from the ethnographic material and an examination of wider Transition materials – in particular, scalar politics and the urban transitions to low carbon societies. The final section discusses the performative nature of the discourses of transition in a city.

Researching transitions in Nottingham

Auto-ethnographic case-study material from the Transition Initiative in Nottingham is addressed, alongside materials draw from the wider Transitions movement. Nottingham was the second city (Bristol was the first) in the United Kingdom to join the Transition network, and provides valuable insights into urban transitions to low carbon societies and adaptations to shocks such as climate change and resource crises. Transition Towns first came to my attention in the spring of 2008 when I attended a talk on peak oil and heard about the Transition Towns initiative that had emerged in Totnes. But my immersion (participatory action research) into the world of Transitioning started with a training event in Nottingham in February 2009, where I was 'trained' (by facilitators from the Transition movement) to initiate a group and was promptly encouraged to work with others on the training course from my neighbourhood to set up a local community group (Transition St Ann's and Sneinton), which has run since February 2009. Furthermore, I have become involved with the wider Transition Nottingham network and am currently a member of a sub-group preparing an Energy Descent Action Plan for Nottingham. I have been very clear to others in all the groups I am involved with that I am an academic and that I research in this area but that I am committed to some of the ideals of Transition Towns and do want to be a participatory action researcher, rather than an observer (Mason and Whitehead, forthcoming, address many of the practical and ethical issues of such research positions). By taking this position, I have built up a wealth of in-depth evidence, including field notes, documents, pictures, minutes of meetings, and so on. This evidence has been analysed using a theoretical framework of performative discourses.

Struggles over meaning are every bit as material as practical struggles (Gramsci 1971). The focus here is not merely upon 'language' but upon *performative* discourses. Hajer defines a discourse as 'a specific ensemble of ideas, concepts, and categorizations that are *produced, reproduced, and transformed* in a particular set of practices and through which meaning is given to physical and social realities' (1995: 44; emphasis added). In many senses, a discourse is 'performative'. Throughout this chapter, it is suggested that Transitions are discursively performed rather than being simply represented through linguistic practice (Knox 2001); thus, it is essential to comment upon the ways in which language has agency, does 'things' and works to achieve outcomes (Parker and Kosofsky Sedgwick 1995). Notions of the *performative function(s) of language* are vital to an understanding of the ways in which linguistic practices are key to the constructing, shaping, moulding, resisting, blending and asserting of Transitioning.

In this sense, Judith Butler's work has been seminal. She has highlighted the connection between power and materialisation. She suggests that by destabilising the binaries between what (which discourse) has been included/excluded (normalised/abnormalised), we can observe the politics of how and why 'something' has come to be excluded, as that which has been excluded is not separate or existing outside, but is also produced by 'the mode of exclusion' (Butler 1993: 35, 39). Within her work on gender, Butler highlights how the repeated use of the same items of rhetoric leads to an accumulation of citational power through which ideas become stabilised (normalised). She calls this process 'citationality' (ibid.: 12). This is of importance to the chapter's focus upon the popularity (the included/the stable/the normalised) of particular constructions of Transition over others (the excluded/the fluid/the subversive).

Discourses of any concept, then, should be analysed in terms of the *claims* themselves, the *claims makers* and the *claims-making process* (Best 1987; Hannigan 1995). If terms acquire meaning from their function within culture (see Wittgenstein 1953), then the main questions to be asked of any 'truth' claims are: 'how do they function, in which rituals are they essential, what activities are facilitated and what impeded, who is harmed and who gains by such claims?' (Gergen 1994: 53). By examining the rhetoric of a truth claim, it is possible to explore the social power and legitimacy of that discourse (Demeritt 1998; Butler 1993). Examining the claims makers and the processes (structures of legitimacy; see Hattingh Smith 2005) via which they make their claims allows comment to be made about the social capital involved in the constructing, shaping, moulding, resisting, blending and asserting of Transitioning. Importantly, some actors will have access to discourses and discursive spaces that have a more significant impact (social power and legitimacy) than others, and this is perhaps dependent upon issues of social capital. For Putnam (1995: 66), social capital 'refers to features of social organization such as networks, norms, and social trust that facilitate coordination and cooperation for mutual benefit', and the stronger the networks of civic engagement, the stronger the abilities are of such networks to coordinate and communicate actions. In particular, Pelling and High (2005: 309) ask, '[W]hat is it about the internal working of communities and organisations that may determine their choices of adaptive strategy?' They further suggest that we can learn from 'ethnographic work into the nature of collective identity and action' (ibid.: 317), and this is a key aim of this chapter. Indeed, identifying the *claims makers* (the key actors and their capacities), their *claims* (key discourses and rhetorical devices) and the *claims-making process* (the discursive spaces and networks of civic engagement) within the Transition movement can provide key lessons for galvanising public support and moving forward key notions of urban Transitioning, adaptation and the fostering of resilience.

Background to Transition Towns and the performative discourses of Transitioning

In so many ways, Transition Towns, the movement and the culture, *are* Rob Hopkins.[3] He is very much the key *claims maker*, and this section focuses largely upon his 'journey' to becoming the co-founder of the Transition Network and founder of the first Transition Town in Totnes, Devon. This includes pointing to some of the key discourses he draws upon, and his and his colleagues' capacities to utilise discursive spaces/networks and circulate subversive/challenging new discourses for action/Transitioning.

Rob is rooted in permaculture, having undertaken a permaculture design course in 1992. He then focused his undergraduate dissertation on the topic while at UWE Bristol, and subsequently moved his family to Ireland to teach permaculture and set up an eco-village development based upon the principles of permaculture. Permaculture is very much modelled on a whole-systems approach to meeting needs; it is a

> process of looking at the whole, seeing what connections can be changed so that the place can work more harmoniously. . . . Although permaculture started out as **perma**nent agri**culture**, the principles on which it is based can be applied to anything we do, and now it is thought of as **perma**nent **culture**. It has grown to include: building, town planning, water supply and purification, and even commercial and financial systems. It has been described as 'designing sustainable human habitats'.
>
> (Whitefield 2000: 3–4; original emphasis)

While teaching permaculture in Ireland (during 2004), Rob first learned about peak oil and was highly influenced by the petroleum geologist Colin Campbell, who has highlighted the end of cheap oil and its consequences for society: 'The world is not running out of oil – at least not yet. What our society does face, and soon, is the end of the abundant and cheap oil on which all industrial nations depend' (Campbell and Laherrere 1998: 83).

Rob has suggested that the issues of peak oil and climate change can be viewed as the 'hydrocarbon twins', given their interconnections and requirements for a holistic approach (Hopkins 2008: 18). While in Ireland, Rob organised a conference in July 2005 ('Fuelling the Future') which included the world's leading speakers on peak oil, such as Colin Campbell, Richard Heinberg, Richard Douthwaite and David Holmgren, and worked with students on his course to produce the first Energy Descent Action Plan for a town (Kinsale, Ireland) – a plan for Transitioning the town from dependency on fossil fuels.

In September 2005, Rob moved to Totnes, Devon, to start a PhD at Plymouth University (focusing on Energy Descent Action Plans). In Totnes, Rob joined forces with many people and started a series of film screenings and discussions around the topics of peak oil and climate change. In September 2006, Transition Town Totnes was launched at an 'Official Unleashing', and in 2008 Rob published *The Transition Handbook: From Oil Dependency to Local Resilience* (Hopkins 2008).

For many, the handbook has become the bible of Transitioning, and should be set alongside the Transition Network site[4] (established in spring 2007), which also features the *Transition Initiatives Primer* and outlines the Transition mission: 'to inspire, inform, support, network and train communities as they consider, adopt and implement a Transition Initiative. We're building a range of materials, training courses, events, tools & techniques, resources and a general support capability to help these communities' (Brangwyn and Hopkins 2008: 3). A Transition Initiative is likely to be undertaken by a community

> in response to the twin pressures of Peak Oil and Climate Change[;] some pioneering communities in the UK, Ireland and beyond are taking an integrated and inclusive approach to reduce their carbon footprint and increase their ability to withstand the fundamental shift that will accompany Peak Oil.
>
> (ibid.)

A Transition Initiative is based on four assumptions (Figure 11.1) and involves a twelve-step process, or journey (Figure 11.2). There are several, often

1. That life with dramatically lower energy consumption is inevitable, and that it's better to **plan** for it than be taken by surprise.
2. That our settlements and communities presently lack the **resilience** to enable them to weather the severe energy shocks that will accompany peak oil.
3. That we have to **act** collectively, and we have to act now.
4. That by unleashing the collective genius of those around us to creatively and proactively design our **energy descent**, we can build ways of living that are more connected, more enriching and that recognise the biological limits of our planet.

Figure 11.1 The four assumptions of Transition Towns.

Source: Adapted from Hopkins (2008: 134).

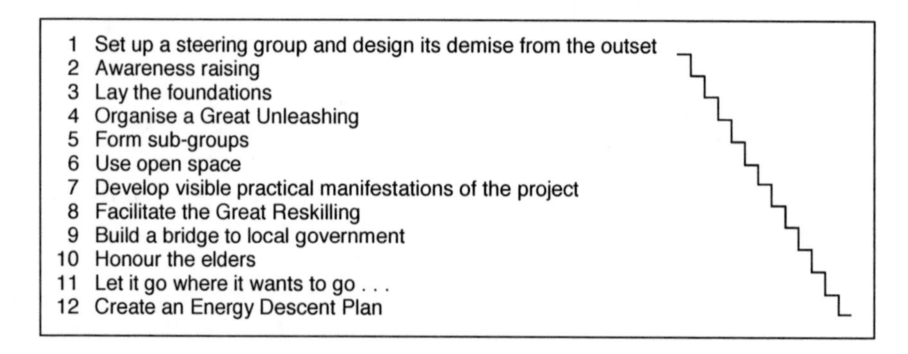

Figure 11.2 Twelve key steps to embarking on the Transition journey.
Source: Adapted from Transition Towns (2009b).

interrelated, key performative discourses (*claims*) in play within these ascribed assumptions and steps, most notably those of:

- *resilience* – which underpins the movement; many of the other discourses stem from this (see below);
- *permaculture* – holism, systems and localisation of food production and economies;
- *community* – inclusivity, social capacity and level or scale of approach-localisation;
- *infectiousness* – methods of communication, the spread of the movement and its 'happy' nature;
- *Transitioning* – skilling up to powering down: changing attitudes and behaviours, capacity building and bringing about collective action.

There are now nearly 300 Transition Initiatives in countries across the world. and for all of them Rob's journey (and those of others associated with developing Transition Towns) and performative discourses (as expressed through the *Handbook*, subsequent publications,[5] the website, online videos and materials, and talks by Rob himself and/or those he works with) are key to their initiation and development. For instance, in order to become an 'official' Transition Initiative there are sixteen criteria to be met (see Figure 11.3). These criteria are a blend of knowledge, skills, communication, working parameters and constitutional aims, but are essentially rooted in the discourses of resilience, localisation, permaculture and notions of community.

Criterion 3 of 'becoming official' requires at least two people from an initiating group to attend a two-day training event (see Figure 11.3). These events originally took place in Totnes but have now been rolled out across the United Kingdom and further afield; they are run by facilitators who are sanctioned by the Transition Network and include[6] a focus upon the context of peak oil and climate change; the

1. An understanding of peak oil and climate change as twin drivers (to be written into constitution or governing documents).
2. A group of four or five people willing to step into leadership roles (not just the boundless enthusiasm of a single person).
3. **At least two people from the core team willing to attend an initial two-day training course. Initially these will be in Totnes and over time we'll roll them out to other areas as well, including internationally.**
4. A potentially strong connection to the local council.
5. An initial understanding of the twelve steps to becoming a Transition Town.
6. A commitment to ask for help when needed.
7. A commitment to regularly update your Transition Initiative web presence.
8. A commitment to make periodic contributions to the Transition Towns blog.
9. A commitment, once you're into the Transition, for your group to give at least two presentations to other communities (in the vicinity) that are considering embarking on this journey.
10. A commitment to network with other Transition Towns.
11. A commitment to work cooperatively with neighbouring Transition Towns.
12. Minimal conflicts of interests in the core team.
13. A commitment to work with the Transition Network re grant applications for funding from national grant-giving bodies. Your own local trusts are yours to deal with as appropriate.
14. A commitment to strive for inclusivity across your entire initiative. We're aware that we need to strengthen this point in response to concerns about extreme political groups becoming involved in transition initiatives.
15. **A recognition that although your entire county or district may need to go through Transition, the first place for you to start is in your local community. It may be that eventually the number of transitioning communities in your area will warrant some central group to help provide local support, but this will emerge over time, rather than be imposed. (This point was inserted in response to the several instances of people rushing off to transition their entire county/region rather than their local community.) Further criteria apply to initiating/coordinating hubs – these can be discussed person to person.**
16. And finally, we recommend that at least one person on the core team should have attended a permaculture design course . . . it really does seem to make a difference.

Figure 11.3 Criteria for becoming a Transition Town.

Source: Adapted and abridged from Transition Towns (2009a).

principles of Transition; visioning exercises; workshops on how to raise awareness of Transition issues; how to establish and maintain an initiating group; a focus upon the psychology of change, transition and addiction; Energy Descent Action Plans; and a 'meeting the descendants' exercise. The course costs around £100 for each person. There are bursaries available (funded from donations by other members of Transition Towns); however, charging for the training will almost inevitably exclude some members of the community. Furthermore, those 'trained' are likely to be seen by other community or group members as having more 'social capital' or discursive power, with an ability to draw upon legitimised discourses of Transition and stabilise or normalise such discourses through citationality. By fulfilling the criteria set out in Figure 11.3, groups 'signing up' to become 'official' are entering into the *claims-making process* (structures of legitimacy) and are agreeing to become *claims makers* who use legitimised *claims* of transition in certain rituals.

Transitions and resilience: an extraordinary renaissance?

> *Central to this book* [*The Transition Handbook*] *is the concept of resilience* . . . the ability of a system, from individual people to whole economies, to hold together and maintain their ability to function in the face of change and shocks from the outside.
>
> (Hopkins 2008: 11; emphasis added)

The concept of resilience forms a very important part of the Transition movement and requires attention here before we can move on to explore how discourses of Transition are being constructed in Nottingham. Resilience has a multilayered history which can see it stretched from its original meaning in ecology (Folke 2006; Gallopin 2006), and has been of particular interest in work addressing complex social environmental systems (McDaniels *et al.* 2008). Indeed, the concepts of resilience, vulnerability and adaptation are seen as increasingly important in the study and responses to environmental change and resource crises (such as climate change and peak oil) (see Janssen and Ostrom 2006; Folke 2006 for in-depth discussions of these concepts). In social environmental systems, the concept of resilience has considerable power to focus on the capacity of humans to anticipate and plan for future events, with the ecosystem in mind. The Resilience Alliance suggests that

> 'Resilience' as applied to ecosystems, or to integrated systems of people and the natural environment, has three defining characteristics:
>
> - The amount of change the system can undergo and still retain the same controls on function and structure
> - The degree to which the system is capable of self-organization
> - The ability to build and increase the capacity for learning and adaptation.
>
> (Resilience Alliance 2009)

In many ways, the Transition movement addresses the last two points by initially drawing on the discourse of resilience to galvanise public support and action to build social capital, and even infrastructure, for adaptation. For instance, resilience 'in the context of communities and settlements . . . refers to their ability to not collapse at the first sight of oil or food shortages, and their ability to respond with adaptability to disturbance' (Hopkins 2008: 54).

Furthermore, in the *Handbook*, Rob draws on work from a number of key authors to stress that the concept of resilience can take us further than the concept of sustainable development when considering peak oil. He suggests (drawing on Levin 1999) that there are three key 'ingredients' to resilience: *diversity*, *modularity* and *tightness* of feedbacks. In terms of diversity, Hopkins suggests that the resilience of a system (whether it be made of elements such as people, species, businesses, etc.) comes not only from the number of species in the system but also from the connectivity between them – and points specifically to our over-reliance on monoculture and the loss of species diversity, and therefore the associated need for diversity of land use. Furthermore, diversity can also refer to the diversity between systems, and Hopkins calls for place-specific, community-led, small-scale solutions:

> The exact set of solutions that will work in one place will not necessarily work in other places: each community will assemble its own solutions, responses and tools. . . . [I]t makes top-down solutions almost redundant . . . resilience-building is about working on small changes to lots of niches in the place, making lots of small interventions rather than a few large ones.
>
> (2008: 55)

In terms of governance, Hopkins stresses that both top-down and bottom-up approaches are required, but that 'we don't need to wait. . . . We can't wait for governments to take the lead . . . we can do a huge amount without government, but we can also do a great deal more with them' (ibid.: 76–77).

The second key ingredient of resilience for Hopkins (drawing on Walker and Salt 2006) is modularity, which refers to the means via which elements in a system are linked. For instance, a system with too many links (over-networked) has high connectivity, which means that shocks in one area can travel quickly through the system (indeed, the credit crunch of the 2000s is a very clear example of this). For Hopkins (2008: 56), 'A more modular structure means that the parts of the system can more effectively self-organise in the event of a shock', and he stresses the need for local food systems, local investment models (such as the Totnes Pound[7]) and general localisation of systems and self-sufficiency. He contends that, third, more localised systems have a tightness of feedback, whereby the impacts of an action can be seen and felt at a local level and therefore acted upon more expediently. By tightening feedback loops, we 'bring the consequences of our actions closer to home, rather than their being so far from our awareness' (ibid.: 57).

Ultimately, the essence of a Transition Initiative grounded in resilience is described as a renaissance:

> Rebuilding local agriculture and food production, localising energy production, rethinking healthcare, rediscovering local building materials in the context of zero energy building, rethinking how we manage waste, all build resilience and offer the potential for an *extraordinary renaissance* – economic, cultural and spiritual.
>
> (Hopkins 2008: 15; emphasis added)

And perhaps the clues to the speed with which this social movement has been taken up lie within the key performative discourse of resilience and those associated with it, as outlined above. Resilience in an urban context is all the more interesting given the constraints of space (for issues such as localisation of agriculture), multilevel governance and spatial politics, diverse communities and areas of deprivation. The chapter now turns to explore some of these aspects in a brief historical (urban-scale) story of Transition Nottingham, with an examination of how discourses of Transition are being constructed, shaped, moulded, resisted, blended and asserted in Nottingham.

Transition Nottingham: the *city* version of transitioning?

Transition Nottingham (TN) started with a small group of individuals (some from other social movements and activist groups) in the spring of 2008. The group were keen to explore whether the model of transition proposed by Rob Hopkins (as outlined above) could be

> used in a city such as Nottingham with a population 35 times larger than Totnes? It's been said amongst the Transition movement that if you can't make cities resilient and sustainable then there's not much point trying with smaller towns and villages.
>
> (Transition Nottingham 2009a)

Criterion no. 15 in becoming a Transition Town refers to the need for local communities to 'transition' first; that counties or regions are not the best level to work at:

> 15. [A] recognition that although your entire county or district may need to go through transition, the first place for you to start is in your local community. . . . (We've seen several instances of people rushing off to transition their entire county/region rather than their local community, and it doesn't work very well.) In exceptional situations where a 'Local Coordinating Hub' or 'Temporary Initiating Hub' needs to be set up (such as Bristol, Forest of Dean), that hub will have certain responsibilities.
>
> (Transition Towns 2009a)

Furthermore, the criterion hints at issues over city (urban) level deployment of the movement, with the call for a 'hub' to coordinate local neighbourhoods or communities. Indeed, this criterion has subsequently been changed (see Figure 11.3), and the movement is in the process of writing a Transition guide for cities, perhaps illustrating the complex challenge such a scale involves.

In Nottingham, there have been a number of issues with the scale at which to develop the Transition Initiative, who should drive it and which discourses are the most legitimate. In Totnes, the model was for an initiating group to start awareness raising (as in the twelve steps noted in Figure 11.2), but planning its demise from the start, as working groups (on topics such as the local economy, food, housing, energy descent, and so on) emerge and a new 'core' forms. TN started as a small (initiating) group of individuals drawn from across the city. These people were largely from a professional background, with careers in academia, consultancy and medicine, but there were also a number of people who had been keen activists for certain issues for many years, such as cycling or climate change. However, it would be fair to say that the group was rather diverse:

> Our steering group is primarily made up of people who haven't worked together previously and this has meant that it's taken a while to begin to feel that we're a team and not just a loose assortment of people interested in the process. It's also meant that we've had to put a lot of energy into making sure that we have good communications and that jobs get done.
>
> (Transition Nottingham 2009a)

This quotation, taken from the TN website, perhaps hints at the issues with communication and the slow progress TN made in the early stages. However, early on in the TN process the group felt that

> [d]ue to its size . . . Nottingham had to be broken down into smaller neighbourhoods in order to carry out the process. . . . We now have seven neighbourhood groups in the city and are planning events to encourage the other areas to begin the process . . . it's when neighbours get together and decide they want to do something to make their area better, and more resilient, that the power of the transition idea becomes evident.
>
> (Transition Nottingham 2009b)

The localisation of Transition is a key discourse of the Transition movement (as noted earlier), particularly in relation to resilience, permaculture and community. Indeed, as part of the process of establishing Transition Initiatives in neighbourhoods, Rob Hopkins was invited to Nottingham to talk about Transitions and generate interest in the issues and process. After his visit, several neighbourhood groupings, in largely middle-class neighbourhoods with higher degrees of social capital, were established and a central 'hub' was delineated, which took a co-ordinating role (see Figure 11.4) and organised city-wide awareness-raising events (such as the Midsummer Energy Festival; see Figure 11.5). However, the actual role of the 'hub' and the local groups became increasingly ambiguous, with members of local groups feeling that the 'hub' was a little too paternalistic and sometimes slow and unresponsive.[8] The classic issues of community engagement came into play (see Smith 2008), with some (hub) group members having 'hidden agendas'; thus, suspicion and a lack of trust became rife across the hub and local groups. Local groups became more distant and communicated less and less with each other and the hub. Furthermore, some members of the hub group were speaking to wider bodies, such as the city council, on behalf of TN but at times without there being a consensus among the wider TN membership, and particularly the local groups. Arguments at the hub meetings seemed commonplace, which in turn was off-putting for local groups attending the meetings.

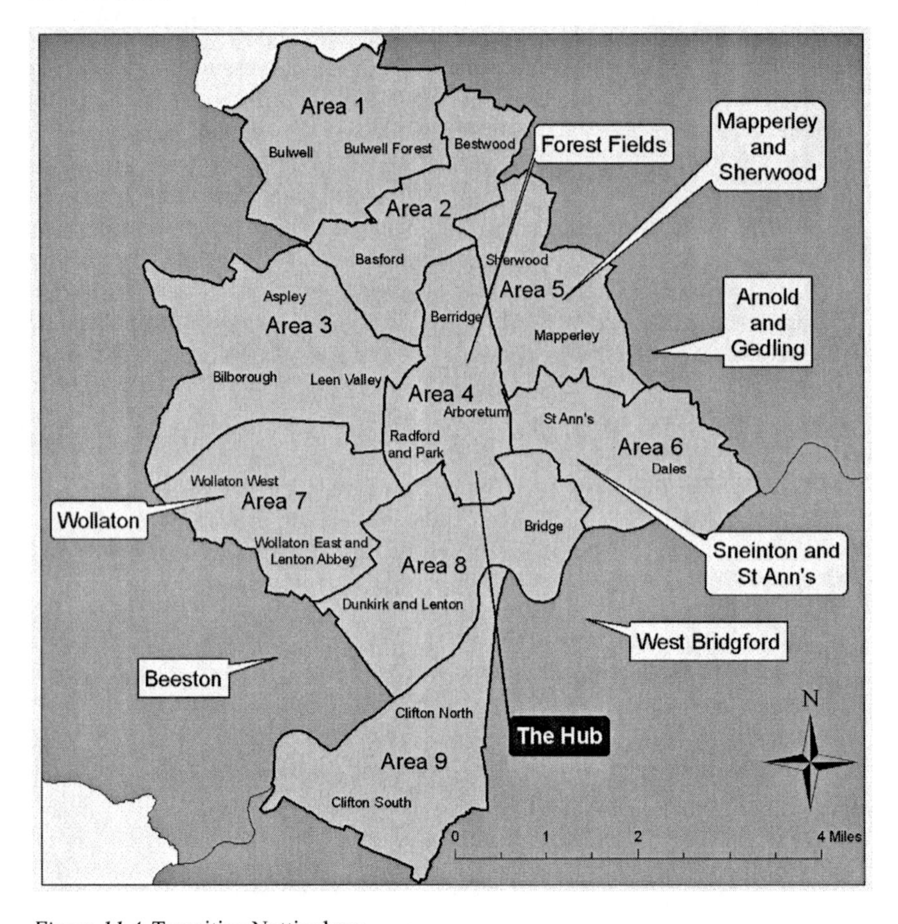

Figure 11.4 Transition Nottingham.

Source: P. Pierce (2010), containing Ordnance Survey data © Crown copyright and database right (2010) adapted.

In order to address these issues and move forward, the 'hub' organised a day event to explore 'unleashing the potential of local groups', where open-space technology was used to facilitate discussion and decision making. At this event, the relationship of the local groups to the hub was debated and the key messages included the following:

- problematics of *doing* Transition in cities: 'We are work in progress compared to Totnes';
- that the hub may need to decline and evolve;
- that communication between all the local groups and the hub needs to be worked on; and
- that a hub of sorts is necessary as a contact point (particularly for bodies such as the city council); also, some issues are bigger than the local groups, such as

PEDAL POWERED FUN - SMOOTHIE MAKER - PLAYSTATION

STILT WALKING - DRUMMING LIVE MUSIC & ENTERTAINMENT

D.I.Y ENERGY GENERATION - WIND TURBINES - SOLAR HOT WATER

HAVE FUN WITH SCIENCE - CHILDRENS ACTIVITIES

LOCAL HOT FOOD PRODUCERS FROM NOTTINGHAMSHIRE

ECO-REFIT YOUR HOME ON A BUDGET - FILMS & INFORMATION

LOTS MORE...

FREE

COME AND EXPLORE ENERGY GENERATION IN ALL OF ITS FORMS

WWW.TRANSITIONNOTTINGHAM.ORG.UK/MIDSUMMERENERGY

Supported by The Nottingham Energy Partnership

Figure 11.5 Midsummer Energy event poster.

Source: Nichola Musgrove.

> Energy Descent Action Planning (EDAP); and because Transition Nottingham exists as a legal body with charitable status, the local groups are not as formalised and cannot apply for funding without the aid of TN.

At a subsequent meeting, a new structure for TN and the local groups was agreed. The hub evolved into a 'support group' for the local groups and wider membership of TN. This support group has specific roles for individuals such as secretary, local groups coordinator, events manager, communications, treasurer and EDAP representative. Big decisions about the future of TN will be made at future open-space meetings where all members of TN will be involved.

This structure has been in place since July 2009 and at the time of writing (February 2010), issues of using the Transition Town performative discourses in Nottingham are still proving problematic. Membership of the local groups has waned and hence their activities in terms of awareness raising have also dipped. A Nottingham Transition Market has been established on a bimonthly basis which allows for the sale of locally produced food and products and sets aside one stall for a local group to run skills workshops. But at present, only one local group has used this stall; it has been difficult to get other local groups involved, possibly owing to the perception that the market is geographically located in that local group's 'area'. Furthermore, work has started on EDAP, but getting a group of people together to work on this has been challenging. The wider membership of TN has been reluctant to attend meetings associated with EDAP, and in an attempt to start the process, people from 'outside' the membership have been 'drafted' in. This has led to further resistance, with arguments over the purpose and style of EDAP becoming commonplace as 'newcomers' question and challenge the key discourses of Transition Towns. In discussions with a number of members of TN, it is clear that the process of performing Transition discourses and building resilience in Nottingham has encountered a number of social power and legitimacy issues that have hampered its progress of late.

Urban transitions and performative discourses

This brief overview of *attempts* to perform Transition Town discourses in the city of Nottingham illustrates the ways in which such discourses are being constructed, shaped, moulded, resisted, blended and asserted in Nottingham. It is clear that the key discourses of the Transition Town movement were reasonably successful at first, with a groundswell of interest in the movement after Rob Hopkins' talk and the first Transition Cities Conference in November 2008 being held in Nottingham. However, it has been difficult for TN to follow (or even deploy) Transition's performative discourses for a number of reasons, but let us frame them in terms of a discursive analysis and focus upon the *claims makers*, the *claims* and the *claims-making process*.

First, this chapter has highlighted that Rob Hopkins is a key *claims maker* in the Transition Town movement, and he in turn draws upon a blend of ideas and work from others such as those addressing peak oil, addiction, resilience, localisation and communities. I have met people within Nottingham who are keen on the general notions of Transition Towns but find the process a little too much like a 'cult'; while others, generally those within TN, had a tendency to view Rob as having a great deal of wisdom and authority. Indeed, in the foreword to *The Transition Handbook*, Richard Heinberg (a key writer and activist on peak oil) is keen to stress that 'perhaps the buzz emanates partly from Rob's own contagious optimism. But this is no personality cult, since Hopkins is quick to cede the limelight to others whenever possible' (quoted in Hopkins 2008: 9).

While not a 'personality cult', the movement does have a tendency to *feel* like a cult, with its performative discourses grounded in sociological and psychological notions that explore, and indeed aim to change, behaviours. In particular, the use of a twelve-step process has distinct similarities to the Alcoholics Anonymous programme, which Alexander and Rollins (1984) have described as 'the unseen cult'. Yet such performative discourses are being resisted and moulded by some claims makers within Nottingham. The initial group who formed TN had a considerable amount of social capital in terms of their socio-economic backgrounds and links to other networks of community and governance within Nottingham, yet as the 'local' groups formed, this social capital became cited there, and those *not* trained in the discourses of Transition challenged the original TN structure and discourses. This led to a moulding of the way in which Transition is being deployed in the city. The local groups are now more likely to be mobilising discourses of Transition than the hub, and these discourses of transition are perhaps not in line with the original *brand* of Transition Towns – given that not all the local groups have people who have been 'trained'. The discourses may well be more grassroots, but this then gives rise to problems of how such discourses may be scaled up to wider urban transitions.

Second, then, the claims of the Transition movement have been moulded too. There is still a focus upon the key discourses of resilience, permaculture, community, infectiousness and transitioning, but different local groups are interpreting these in their own ways. And it is becoming increasingly problematic to address the city-wide issues such as devising an Energy Descent Action Plan when local groups are building their own resilience and becoming increasingly inward-looking. The bottom-up, small-scale adaptations that Hopkins calls for to build resilience have perhaps strengthened this insularity – or even put some people off, as the movement does not attempt to tackle wider socio-economic structures, such as a macroeconomic system based on economic growth, and can be described, to a certain extent, as depoliticised (Trapese 2008).

Indeed, the *claims-making* process, within which certain discursive structures are legitimised over others, of Transition Towns does call for an engagement with local government (see Figure 11.2), yet it attempts to remain apolitical and non-campaigning. In Nottingham, this has proved complicated, with the council appearing uncertain on how to engage with TN, and vice versa. That said, the council was the first in the country to pass a motion related to peak oil. The point is perhaps that by trying to act in a depoliticised manner, TN has not engaged in legitimate discursive structures, and thus bottom-up has yet to meet top-down. As Trapese (2008: 7) suggests, 'an agreement "not to rock the boat" will not help TT's long term viability, as it would not mean really changing anything'.

This perhaps helps to explain some of the processes that have taken place in Nottingham, with the complexity of deploying Transition discourses at an urban/city scale combining with a sense of *discursive mutism* whereby the performative discourses have largely failed to perform and have yet to bring about the renaissance in resilience that Hopkins alludes to. That said, I personally will still attempt to remain involved with TN and my own local group, and in a way I too am using my social capital to shape, blend, mould and resist (some) Transition discourses, as I do find that many of these discourses ring true for me and offer a future low carbon society I might like to live in and be part of. However, quite how a city version of Transition can be performed successfully is still open to discursive struggles.

Questions to consider for future urban transitions to low carbon societies

- What is the best scale at which to build resilience and bring about transitions? Small-scale community 'experiments' may yield some interesting pointers for sustainable transitions, but can they be scaled up to wider urban transitions?
- Can we bring about bottom-up transitions without considering wider issues such as rampant consumerism, general adversity to change, and political structures that maintain the status quo?
- What scope is there for a Transition *City* in the light of the complexity of wider socio-economic-technological urban structures? Scaling transitions at an urban scale is conceptually problematic: for instance, which 'community or communities' are we bringing about a renaissance in? Yet cities, as spaces for transition, offer many opportunities, particularly in the light of their nature(s) of diversity, tolerance, creativity, complex networks of social capital and their roles as sites for experimentation and novelty.
- Furthermore, policy frameworks are problematic: can we transition an entire city that sits within a governance regime(s) that is hierarchical and essentially top-down? Do we need further research to consider how best to conceptualise urban transitions? What does a city in transition look like? Can it be 'planned'? Or is it inevitably going to be a dynamic, messy and chaotic process – particularly if it is to be driven by grassroots movements?
- In terms of performative discourses, who is in transition and to what? Is there an agreement on what a 'transition' is? The evidence in Nottingham suggests that there is some consensus but that there are disagreements over the processes of transitioning. Indeed, how will we know when we get there? The Transition Town movement advocates the use of visioning exercises and back-casting to help communities in their Transitions; but in Nottingham, at present, a debate remains unresolved in terms of whether we want a *realistic*

or *idealistic* vision to work towards – or perhaps more importantly the issue is: whose discourse of realistic/idealistic is the most legitimate? Indeed, one person's realistic discourse may be viewed as idealistic by another.

Notes

1 See www.energysavingtrust.co.uk/housingbuildings/localauthorities/NottinghamDeclaration/ for details of the Nottingham Declaration and which local authorities have signed (accessed April 2009).
2 See www.iclei.org/index.php?id=800 (accessed April 2009) for details of this initiative.
3 Information for this section is taken from http://transitionculture.org/about/ (accessed April 2009) and Hopkins (2008).
4 See http://transitiontowns.org/TransitionNetwork/TransitionNetwork (accessed April 2009).
5 Subsequent books published by Green Books on behalf of Transition Towns in 2009 include *Local Food: How to Make It Happen in Your Community* and *The Transition Timeline for a Local, Resilient Future.*
6 This was the content of the training event held in Nottingham in February 2009 and is similar to such events held elsewhere such as Liverpool, Totnes and Derby. The website also lists full details of the training: see http://transitiontowns.org/TransitionNetwork/TransitionTrainingDetail (accessed January 2010).
7 See http://totnes.transitionnetwork.org/totnespound/home (accessed April 2009) for details of the Totnes Pound.
8 Based on my experience of working with Transition Nottingham.

References

Alexander, F. and Rollins, M. (1984) 'Alcoholics Anonymous: The unseen cult', *California Sociologist* 7: 33–48.

Barr, S. (2008) *Environment and Society: Sustainability, Policy and the Citizen*, Aldershot, UK: Ashgate.

Best, J. (1987) 'Rhetoric in claims-making: Constructing the missing children problem', *Social Problems* 34: 101–121.

Betsill, M. and Bulkeley, H. (2007) 'Looking back and thinking ahead: A decade of cities and climate change research', *Local Environment* 12: 447–456.

Bickerstaff, K. and Walker, G. (2003) 'The place(s) of matter: Matter out of place – public understandings of air pollution', *Progress in Human Geography* 27: 45–67.

Brangwyn, B. and Hopkins, R. (2008) 'Transition Initiatives primer'. Online, available at www.paulchefurka.ca/TransitionInitiativesPrimer.pdf (accessed April 2009).

Butler, J. (1993) *Bodies That Matter: On the Discursive Limits of 'Sex'*, London: Routledge.

Campbell, C. J. and Laherrere, J. H. (1998) 'The end of cheap oil', *Scientific American* 288: 78–83.

Demeritt, D. (1998) 'Science, social constructivism and nature', in B. Braun and N. Castree (eds) *Remaking Reality: Nature at the Millennium*, London: Routledge.

Folke, C. (2006) 'Resilience: The emergence of a perspective for social-ecological systems analyses', *Global Environmental Change* 16: 253–267.

Gallopin, G. (2006) 'Linkages between vulnerability, resilience, and adaptive capacity', *Global Environmental Change* 16: 293–303.

Gergen, K. J. (1994) *Realities and Relationships: Soundings in Social Construction*, Cambridge, MA: Harvard University Press.

Giradet, H. (2008) *Cities, People, Planet: Urban Development and Climate Change*, Chichester, UK: John Wiley.

Gramsci, A. (1971) *Selections from the Prison Notebooks*, New York: International Publishers.

Hajer, M. (1995) *The Politics of Environmental Discourse: Ecological Modernization and the Policy Process*, Oxford: Clarendon Press.

Hannigan, J. A. (1995) *Environmental Sociology: A Social Constructionist Perspective*, London: Routledge.

Hattingh Smith, A. (2005) 'Sustainable communities? Lessons from the "sustainable regeneration" of former coalfield communities in East Durham', ESRC/DCLG Postgraduate Research Programme Working Paper 26.

Hodson, M. and Marvin, S. (2009) 'Urban ecological security: A new urban paradigm?', *International Journal of Urban and Regional Research* 33: 193–215.

Hopkins, R. (2008) *The Transition Handbook: From Oil Dependency to Local Resilience*, Totnes, UK: Green Books.

Hopkins, R. and Lipman, P. (2009) *Who We Are and What We Do*, Totnes, UK: Transition Network.

Janssen, M. A. and Ostrom, E. (2006) 'Editorial: Resilience, vulnerability, and adaptation: A cross-cutting theme of the International Human Dimensions Programme on Global Environmental Change, *Global Environmental Change* 16: 237–239.

Knox, D. (2001) 'Doing the Doric: The institutionalization of regional language and cultural change in the north-east of Scotland', *Social & Cultural Geography* 2: 315–331.

Levin, S. A. (1999) *Fragile Dominion: Complexity and the Commons*, Reading, MA: Perseus.

McDaniels, T., Chang, S., Cole, D., Mikawoz, J. and Longstaff, H. (2008) 'Fostering resilience to extreme events within infrastructure systems', *Global Environmental Change* 18: 310–318.

Macnaghten, P. and Urry, J. (1998) *Contested Natures*, London: Sage.

Mason, K. and Whitehead, M. (forthcoming) 'Minding a mendacious methodology: Community-based research in a Transition Town', *Qualitative Research*.

Moser, S. C. and Dilling, L. (eds) (2007) *Creating a Climate for Change: Communicating Climate Change and Facilitating Social Change*, Cambridge: Cambridge University Press.

Parker, A. and Kosofsky Sedgwick, E. (eds) (1995) *Performativity and Performance*, London: Routledge.

Pelling, M. and High, C. (2005) 'Understanding adaptation: What can social capital offer assessments of adaptive capital?', *Global Environmental Change* 15: 308–319.

Putnam, R. (1995) 'Bowling alone', *Journal of Democracy* 6: 65–78.

Putnam, R. (2004) *Democracies in Flux: The Evolution of Social Capital in Contemporary Society*, Oxford: Oxford University Press.

Resilience Alliance (2009) 'Resilience'. Online, available at: www.resalliance.org/576.php (accessed April 2009).

Smith, A. (2008) 'Applying the lessons learnt', *Public Policy and Administration* 23: 145–152.

Transition Nottingham (2009a) 'Starting out'. Online, available at: www.transitionnottingham.org.uk/about_starting.php (accessed April 2009).

Transition Nottingham (2009b) 'Setting up neighbourhoods'. Online, available at: www.transitionnottingham.org.uk/about_neighbourhoods.php (accessed April 2009).

Transition Towns (2009a) 'Becoming official'. Online, available at: www.transition network.org/community/support/becoming-official (accessed April 2009).

Transition Towns (2009b) 'Twelve ingredients'. Online, available at: www.transitionnetwork.org/community/support/12-ingredients (accessed April 2009).

Trapese (2008) *The Rocky Road to a Real Transition: The Transition Towns Movement and What It Means for Social Change*, Trapese Popular Education Collective (www.trapese.org).

Walker, B. and Salt, D. (2006) *Resilience Thinking: Sustaining Ecosystems and People in a Changing World*, Washington, DC: Island Press.

Whitefield, P. (2000) *Permaculture in a Nutshell*, East Meon, Hampshire, UK: Permanent Publications.

Wittgenstein, L. (1953) *Philosophical Investigations*, Oxford: Blackwell.

12 Building liveable cities

Urban Low Impact Developments as low carbon solutions?

Jenny Pickerill

Introduction

This chapter takes the issue of access to low-cost and sustainable housing as a lens through which to explore the possibilities of low carbon transitions in cities. In a normative and broadly supportive way it uses a close examination of low-impact sustainable housing and livelihood projects in rural areas (defined as Low Impact Developments,[1] LIDs) to advocate a social movements approach to shaping transitions in cities. This chapter is empathic and interactive rather than extractive and objective – a reflection of the participatory and embedded way in which the research was undertaken. Such an approach is necessary if we are to assess the normative worth of, and assertively advocate more inclusive and sustainable forms of, transition. However, this stance does not dismiss the importance of critique, and the chapter identifies the constraints and tensions in adopting a social movements approach.

Using LIDs, this chapter seeks to distinguish a certain form of transition – a social movement environmental perspective of change – from an economic or management process. The social movement approach conceives of transition as a powerful form of directed grassroots collective change, often imbued with a hopeful (almost utopian) ambition. Such a perspective might envisage a pre-determined end, but the processes of change and the inclusion of multiple participants are more important than seeking consensus on what this end might be. In contrast, an economic or management approach is very linear in its understanding of social change. It identifies structural causes of problems which can be overcome with financial stimuli or socio-technical innovations. A social movements approach to transition offers a more accurate understanding of the complexity and unknowability of social change, arguing that it is not the management of outcomes in which we should be most interested, but instead the messy processes of grassroots participatory change. Conceptually, this social movement approach builds upon the socio-technical transitions understanding of the importance of innovation. LID is an example of a niche from which social innovation emerges. However, a social movement approach requires us to understand innovation as a more holistic, complex and participatory process. As such, the unfinished nature of LID, its dynamic and fluid experimentation, and grassroots collective action call into question the validity of economic and management approaches to transition.

This chapter begins by outlining briefly how LID is a form of low carbon living and then examines the extent to which it has been put into practice in urban environments. The problems in translating LID, a radical approach to sustainability, into urban contexts are then used as a way to elucidate some of the tensions of attempts at creating low carbon cities and in understanding urban transitions.

LIDs are proliferating across Britain. They are a form of localised eco-development and articulated as an alternative to neo-liberalised forms of living. That is, they critique consumerist capitalist materialism, illustrating ways of living with minimal income, few possessions, limited employment and significantly reduced financial outgoings. We can define LID as an approach that requires changes to our housing, livelihoods and lifestyle, all with an emphasis on ensuring affordability. A LID has a low visual impact by blending with its surroundings, is built from local, recycled or natural materials, is small scale and environmentally efficient. LID has much in common with other eco-communities but its attention to the social (diversity, emotions and needs), the financial (affordability and livelihoods) and its holistic approach sets it apart from other approaches. Many are exemplars of low or zero carbon living (Pickerill and Maxey 2009b). Thus far, the majority of LIDs have been in rural areas using agricultural land to generate income and affordable rural livelihoods. However, the ideas of LIDs can be extended to urban locations. Hence, an examination of LID enables us to explore some of the tensions of transitions within cities – those of scale, temporality, constraints, social inclusion and social justice. LID has its historical roots in quests for survival by the rural poor, the back-to-the-land movement and grassroots sustainability initiatives (such as Agenda 21), but is now (re)cast as part of the solution to our environmental problems (Fairlie 1996; Ward 2002; Pickerill and Maxey 2009b).

This chapter draws on material collected during a thirty-month (2005–2008) empirical participatory research project on LIDs in Britain. In particular, I participated in the development of Lammas Low Impact Initiative Ltd (www.lammas.org.uk) – which was awarded planning permission for nine LID dwellings (smallholdings) in August 2009 on farmland near Glandwr, north Pembrokeshire, Wales. In addition, repeated site visits were conducted to four LIDs (Steward Community Woodland, Hockerton Housing Project, Green Hill (a pseudonym), and Hill Holt Wood), interviews were carried out with their residents, and Brighton Earthship was visited as an example of novel eco-construction.

Low Impact Development and low carbon living

LID is a radical form of housing, livelihood and lifestyle that works in harmony with the landscape and natural world around it (Halfacree 2006). In the past fifteen years, there has been a significant growth in the number of LIDs in Britain (Figure 12.1), with estimates that 10,000 people now live in them (Chapter 7 2003). LID is important as a site of practical innovation and attempts at low carbon living. Its core principles are that in order to reduce our environmental impact we must

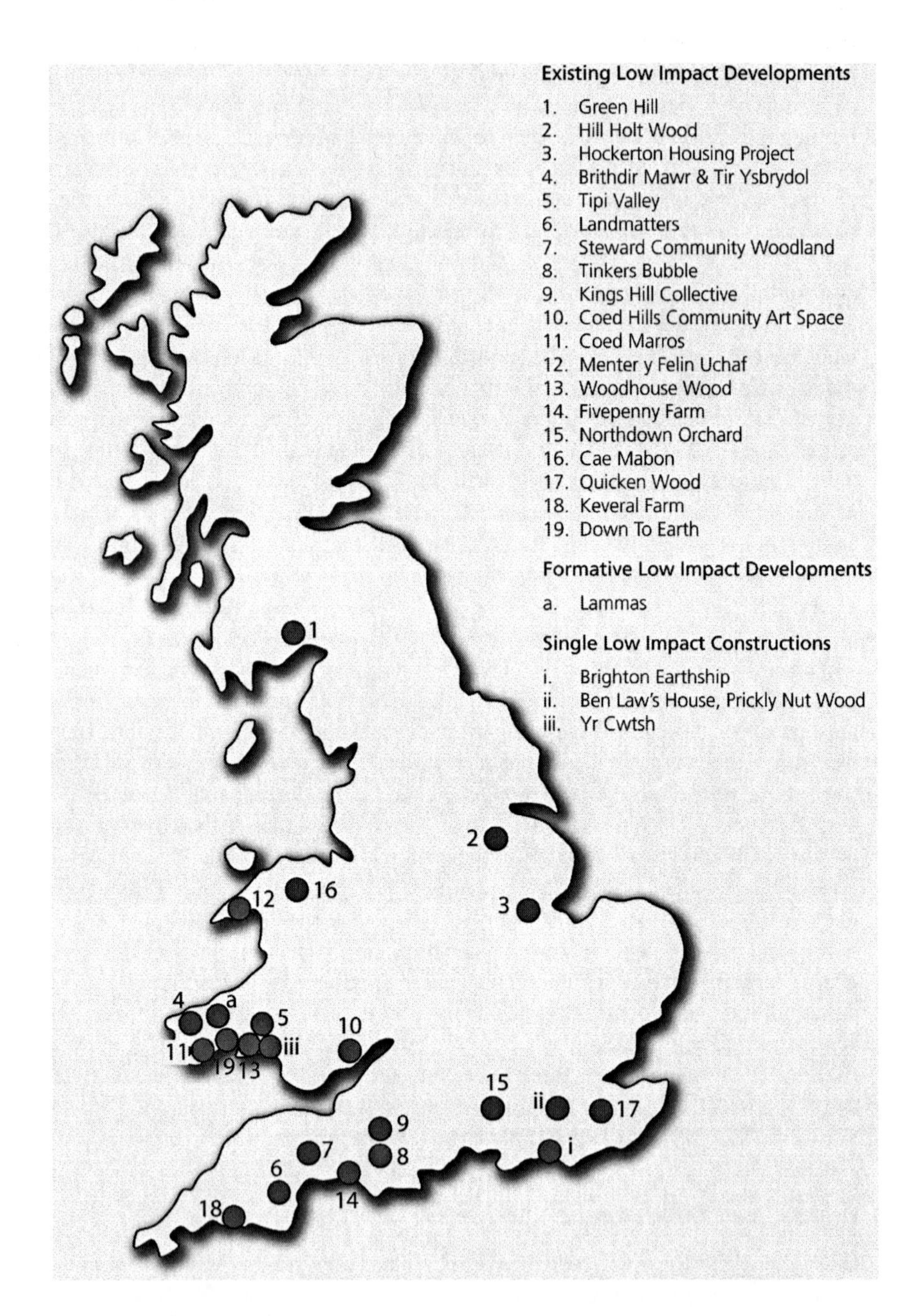

Figure 12.1 Map of some existing and formative Low Impact Developments in Britain, November 2008.

radically redesign our homes, satisfy our needs ourselves or through local provision (be it energy, food, education, etc.), reduce our needs (through reducing consumption), work close to where we live (thus far, mainly involving living off the land as smallholders) and dramatically reduce our travel (for a detailed discussion of these elements, see Pickerill and Maxey 2009b). Buildings use natural local materials wherever possible, renewable energy is generated on-site, waste is processed environmentally (through using reed beds, for example) and little traffic is generated. They are exemplars of low carbon living and meet the criteria of economic, social and environmental sustainability (Dale 2009; UWE and Land Use Consultants 2002). Examples include the 'zero carbon' Hockerton Housing Project and Ben Law's house (Figures 12.2 and 12.3) (for more examples, see Pickerill and Maxey 2009a).

In addition, LIDs emphasise the need for low-cost housing – with some dwellings, such as Tony Wrench's Roundhouse, costing just £3,000 to build (Wrench 2001) – low-cost as regards both set-up and running costs. LID acknowledges that there are significant financial implications in the shift towards low carbon living and a risk that many will be excluded by the high costs of eco-houses and associated technologies. LIDs seek to prove that cost need not be an obstacle. The majority of LIDs thus far have achieved low costs through the use of communal infrastructure and an emphasis upon working together as a community. However, this is not a necessary feature of LIDs, and there are examples, such as Ben Law's house (Figure 12.3), that are single residences.

Figure 12.2 Hockerton Housing Project, Nottinghamshire, April 2007.

Figure 12.3 Ben Law's house, Prickly Nut Wood.
Source: Ben Law.

The evolution of LIDs illustrates several useful insights into attempts at low carbon living and understanding the notion of transition (Pickerill and Maxey 2009b). First, LID is much more than simply a fix to reduce our environmental impact; it is an holistic approach which teaches us that the personal and emotional matter, as well as the practical and technological (Brown and Pickerill 2009). It takes holism – the idea that we need to understand the whole of a system (physical, social, economic and psychological) and that the properties of a system cannot be understood by its component parts alone – as its approach to understanding how humans should interact with the environment. It explicitly builds on the permaculture approach to life, which advocates that we need to pay attention to Earth Care, People Care and Fair Share (Holmgren 2002). This holistic approach advocates that in order to change our relationship to the system (the environment), we must make physical changes (such as growing our own food) *alongside* economic and psychological changes. In other words, without understanding how people shape and stall processes of change, we fail to fully understand LID and environmental transitions more broadly. Part of this involves a concern not just for how we encourage behavioural change, but the personal politics of those changes (Laughton 2008). LID is very hard work, and while much of that work is eased through communal infrastructure, working together creates its own problems: 'just living with each other can be hard – understanding each other, accepting everybody's differences . . . and not being judgemental . . . loving everybody all the time is hard' (May, Green Hill, interview). LID is holistic, and the importance of the personal and community aspects of change are acknowledged. Hence, the more linear approaches to transition do not fit with the experiences of social change that LID has forged. Moreover, any innovation in an LID has to be understood within the holistic framework of other practices, not as an individual practice that can be replicated out of context.

Second, LID is concerned with simplifying our lives and not relying upon technological fixes. However, it also avoids consigning people to a heavy workload without the use of modern tools. For example, one resident is a freelance IT consultant who works from a canvas-covered home in the woods, and another acknowledges, 'I don't ever imagine us to be entirely self-reliant . . . we'd never produce our own wheat or produce our own bread' (Will, Green Hill, interview). Its approach is pragmatic and informed: thus, it advocates that we are conscious of our needs and consciously choose the most appropriate tools, criticising society's relationship with technology as always necessary or the best solution. In understanding transition, LID urges us to acknowledge that simplicity is a powerful force for change and that we need to look beyond technological innovation for solutions.

Third, LID is not a set of rigid principles, but rather a dynamic and evolving set of ideas and inspirations. This adaptability and flexibility is crucial in LID's acknowledging its limitations and adjusting to new contexts (which is why it has potential in urban locations). In many ways, LID is an unfinished experiment. Ensuring that ideas for low carbon living do not become rigid 'blueprints' and leaving some questions and problems open for negotiation is likely to facilitate their adoption by diverse populations. Thus, LID values innovation as an instigator for broader change (just as socio-technical transitions do) but seeks to ensure that we understand these innovations as social constructions (not simply technological fixes) and hence that they continue to evolve and change in an intentionally dynamic way. In other words, it is far more complicated than simply 'scaling up' some aspects of LID practice into the mainstream.

Finally, LIDs are purposely small scale and they raise important questions about how 'localised' low carbon initiatives need to be. Transition can be conceived on a variety of scales, and much of the economic approach is regional or national in its ambitions. LIDs, however, raise important questions about our expectations of the scale and speed of transition.

Urban Low Impact Development?

The lived practice of LID provides valuable evidence for the viability of low carbon living. However, there are currently very few examples of LID in urban[2] contexts. In part, this reflects the roots of LID in land-based livelihoods and smallholder practices:

> We would like to think that this model could be replicated in a forest or a brownfield site. . . . Low Impact Development can happen in an urban space,

> but . . . it is very much geared towards smallholders, and as a land-based thing. I would hope in the future that it can evolve to encompass urban development, but in its present form it's exclusively rural.
>
> (Paul, Lammas, interview)

Rather, there has been partial adoption of some of its components, often divorced from the broader principles of LID – for example, the rise of urban allotment use (and urban permaculture), city farms, eco-buildings, car pooling, collective electricity generation (community wind turbines) and waste reduction (collective composting and recycling) (Girardet 2007; Lovell 2005; Low *et al.* 2005; Rydin *et al.* 2007; Wolch 2007). More promisingly, there are a growing range of collective attempts at more sustainable living in urban spaces, from land squats to co-housing schemes, and many of these share important principles with LIDs.

Pure Genius was a land squat on the 13-acre site of an old gin distillery owned by Guinness in Wandsworth, London (Halfacree 1999). It started on 5 May 1996 and the occupiers were evicted on 15 October that year, just over five months later (Schwarz and Schwarz 1998). It was coordinated by The Land Is Ours as a protest at the lack of power local people had in influencing the planning and land use of cities (Featherstone 1997). The group sought to 'demonstrate green urban living' (Schwarz and Schwarz 1998: 54) by building a temporary eco-community, and they constructed a number of dwellings, including a communal roundhouse, from scrap and reclaimed material (Figures 12.4 and 12.5). Not only did Pure Genius

Figure 12.4 Pure Genius land squat, May 1996.

Source: Keith Halfacree.

Figure 12.5 Pure Genius communal roundhouse, May 1996.
Source: Keith Halfacree.

illustrate that it was possible to build an eco-village quickly in the heart of London – with permaculture gardens, compost, raised beds, irrigation systems and a wind turbine and generator – but the group attempted to deal with issues of accessibility, welcoming homeless people and encouraging them to build their own homes; they 'are low-cost houses for people who would not be housed otherwise' (Monbiot, quoted in Schwarz and Schwarz 1998: 56). The residents were diverse and in flux (ranging in number from 100 down to 50 on the day of eviction, and including local, national and international visitors), gained support from neighbouring residents, and attracted donations of compost, candles and wood from local businesses. The site had problems, in part because of its disorganised organic nature, the flux of a diverse population, and dealing with the reality of inner-city life. The site was always likely to be temporary, but it did engage with issues of housing, work, lifestyle and affordability, and served as inspiration for many activists now living in LIDs: 'I remember being quite inspired by the Guinness project down in London' (May, Green Hill, interview).

Other urban projects have been more permanent, but in being so have tended to miss one or more of the core components of the LID approach. BedZED (Beddington Zero Energy Development) was completed in March 2002 in Wallington, south London (Figures 12.6 and 12.7). Designed by Bill Dunster, it was the first large-scale (eighty-two homes) urban eco-development in Britain (Lazarus 2003). In particular, it aimed to be a 'zero fossil energy development' – producing as much energy as it consumed, and using low-impact materials for

Figure 12.6 BedZED, March 2009.
Source: BioRegional Development Group, BedZED Centre.

Figure 12.7 BedZED, May 2008.
Source: BioRegional Development Group, BedZED Centre.

construction, sourced from within 35 miles of the site (Twinn 2003). In addition to its eco-design, residents are encouraged to adjust their lifestyle by joining a car club, use roof gardens (Figure 12.8) and rent workspaces on site (Hardy 2004). It

> achieves an overall density of 50 dwellings per hectare, 120 workspaces per hectare, and over 4000m^2 of green open space per hectare. At these densities nearly three million homes could be provided on existing brownfield land, at the same time as providing all the workspaces needed for the occupants.
>
> (BRESCU 2002: 9)

However, there are few data available on the extent to which residents have adjusted their lifestyle to fit BedZED's goals; moreover, although the homes are of mixed tenure and 'two thirds affordable or social housing' (BRESCU 2002: 3), it is unclear how this affordability might be protected or whether social housing really represents low-cost housing. Twinn notes early on that 'interest has continued to increase so that the homes command a significant premium above market rates' (2003: 16). Finally, several of the communal pieces of infrastructure, such as the combined heat and power system (gasifier) and the reed-bed sewage filtration system of recycling waste water, have suffered from breakdown and high maintenance costs, which means that they are no longer used (Dawson 2006). While much of BedZED exemplifies an LID approach, especially the combining of home and workspace, and attempts at affordability, the failure of some of the

Figure 12.8 Roof garden at BedZED, July 2009.

Source: BioRegional Development Group, BedZED Centre.

communal infrastructure raises questions about its long-term resilience. It is unclear whether the fact that this was a developer-led project, rather than built and designed by residents themselves (as in most LIDs), has any role to play in that resilience. It does diverge from the model of LID as grassroots collective change, although, as Smith (2007) identifies, BedZED could act as an 'intermediate' niche between the radical innovation of LIDs and the regime, enabling the ideas of LIDs to reach a broader audience.

Although co-housing is not necessarily environmental in design, Springhill Cohousing (www.springhillcohousing.com) in central Stroud was completed in 2003 and has many green features, such as highly insulated dwellings and photovoltaic roof tiles (Heeks 2007; Meltzer 2005). It is the first co-housing urban project in Britain. It has thirty-four units (flats and houses) and a communal building and shared facilities. In line with other co-housing, it is participatory in its design and management. However, it only has one workplace studio, and although it has 'affordable units', they are currently selling at average prices for the area. Questions have also been raised as to the lack of diversity of its population (Newton 2008).

In Berlin, Ufa-Fabrik is a long-running commune founded in 1979 as the result of a land squat on the site of the Universal Film Studio film processing factory. It is unusual for its urban location, has many eco-features including wind turbines, photovoltaic panels and a reed bed system, and a population of around 50 (Miles 2008). It is a cultural space as well as a residential community, with numerous workshops and rehearsal spaces, and a café; arts events are regularly held there. It employs around 100 people and has its own primary school.

There are also promising new projects in the pipeline. In Brighton, planning permission has been granted for a sixteen-unit earthship development, called Lizard. Constructed from old tyres and other waste, earthships[3] are carbon neutral. This will be the first time they have been used as residences in Britain, and to such a dense design (Hewitt and Telfer 2007). However, they are unlikely to be low cost; they will sell for market rates. Activists continue to reclaim city spaces and establish eco-villages, such as Kew Bridge in London during the summer of 2009 (Anon., 2009). Finally, Lilac (Low Impact Living Affordable Community) (www.lilac.coop) is based in Leeds and intends to create a low-impact urban low-cost community based on the principles of co-housing. It intends to be highly environmental in its design, but does not at present include workspaces in its plans.

These examples illustrate the feasibility of various aspects of the LID approach, albeit with varying success. However, an LID approach asserts that in order to create low carbon futures we must adjust our style of housing, our work practices and our personal behaviour, and ensure that these choices are low cost. Thus, LID is an holistic approach, not one that will necessarily work if only certain elements are 'cherry-picked'; and there remain too few working examples in urban locations.

Building blocks of a low carbon future

Urban LID answers a number of ongoing arguments about what a 'sustainable city' should look like (Girardet 2007). Our future urban spaces need to be better designed to reduce our collective environmental impact, but they also need to be accessible to a diverse population. High-density LIDs could also take more advantage of the benefits of communal infrastructure (energy generation and community gardens), additional employment opportunities, existing community networks, reduced transport pressures (yet retaining mobility), greater education choices and the integrated cultural opportunities of urban locations. However, examination of the progress in developing urban LIDs enables us to identify several constraints and tensions in building low carbon futures. First, LID is purposefully small scale. The larger projects discussed above were developer led rather than participatory, and this potentially limited commitment to solving problems with communal infrastructure. Existing projects are all small pockets of change within broader unsustainable cities, and it is clear that such projects need to be better networked and actively promoted at regional and national levels if they are to have a broader impact. In this way, there is potential in coordinating small-scale projects to increase their political and practical impact, but also for more critical discussion as to both what constitutes 'small scale' and the implications of localisation or bioregional approaches.

The question of how much space an individual needs to provide their own food remains debatable; one site, Lammas considered,

> was on the edge of an industrial area, but that wasn't enough land really to allow people to be self-sufficient, or even approaching self-sufficiency, so that was kind of why we discounted that. The land-based nature of what we're doing means that a rural location is needed.
>
> (Larch, Lammas, interview)

An urban context does provoke creative permaculture solutions to intensive cultivation, but few urban projects have attempted self-sufficiency in this way, and large cities remain reliant on vast hinterlands and the unsustainable transport miles involved in food delivery (Holmgren 2002).

Second, a core principle of LID is that work and residence are combined to provide local livelihoods. The provision of workspace has been accommodated by many projects already, but issues remain about the space required if more land-based income-generating projects were to be adopted from existing LID experiments. Certainly the requirement to think more creatively about meeting household needs from on-site activity would reduce transport pressures and increase neighbourhood interaction, and, as was noted previously, urban locations enable a far broader range of localised employment opportunities than do rural sites.

Third, LID takes a long time – not just in physically finding sites and self-building dwellings but, perhaps more importantly, in encouraging others to adopt quite radical ideas, and the emotions and politics of working collectively (though not all LIDs require collective living). Our own processes of change may not be quick enough to accommodate our changing climatic conditions. Indeed, despite the vast experience of intensional living, there remains a tendency in LID to favour practical work rather than people or emotional work.

Fourth, there is a tension in the importance of place to such projects. LID emerged from a specifically rural context, and the suggestion of its transference to urban locations raises important questions about how ideas travel and adapt. Is there something unique about the urban context, and does it matter where ideas first come from? Ideas, especially green ideas, are often imported, but they take time to be adjusted to suit new places. For example, the Brighton Earthship was based on the principles of the earthships in the New Mexico deserts of the United States, and in hindsight required additional heating technologies to suit the British climate (the coordinator lamented the lack of ground-source heat pumps) (Reynolds 1990). Future versions, such as at The Lizard, have such additional features included. However, the notion of a rural idyll is incredibly powerful for many, especially among those seeking to create a more sustainable lifestyle: ‘Doing it in a city didn’t appeal to me at all; I definitely wanted to be out in the wilderness a little bit’ (Jo, Green Hill, interview). Although there are groups that seek to celebrate and reclaim the ‘wastelands’ of urban spaces (for example, the City Repair project; www.cityrepair.org), valuing urban places as environmentally significant and biodiverse can be a difficult task.

Fifth, poor governance structures significantly constrain the possibilities of urban LIDs. The ongoing struggles for rural projects to secure planning permission signal the failure of current planning systems to enable, let alone encourage, radical enough environmental solutions to be practised (Fairlie 2003; Corkindale 2007). The more successful urban LID attempts, such as BedZED and Springhill, were led either by developers or by individuals who had the skills and time to work through the system. The only way that future urban LID projects will occur is with greater provision of land by councils (for example, BedZED was partly secured by the Peabody Trust from the council for a lower rate in return for providing social housing) and changes in planning regulations (Kanaley 2002; Hall and Purchase 2006). The UK planning guidance PPS4, released in December 2009, decisively works against approval of LIDs by prioritising the approval of new rural commercial businesses (measured by profitability), while denying LIDs because they do not add significantly to local economies.

Finally, LID asserts that environmental problems cannot be solved by environmental solutions alone, and that only a holistic approach which attends to social justice issues of social inclusion, poverty and accessibility will facilitate low carbon futures. The issue of cost has barely begun to be tackled, with apparently ‘affordable’ units being far out of the financial reach of many. Yet LID offers the potential to radically reduce construction, maintenance and energy costs for many residents. The issue of diversity of residents, whether it be across race, class, income, educational attainment or many other divisions, remains problematic. Many rural LID residents lamented their homogeneity, but for Pure Genius, diversity was a particular friction. Others hope that an urban location would enable far more integration and diversity: ‘It could work in more of an urban space, in a brownfield site . . . having more people involved and creating a true eco-village of more people’ (May, Green Hill, interview).

Processes of urban transition

If LID is a useful lens through which to understand some of the ongoing tensions in attempts to create low carbon cities, it can also be instructive when considering debates on the processes of urban transition per se. The concept of LID, and how it has been practised thus far in Britain, is rooted in multifaceted pragmatism. On the one hand, it is a flexible, evolving, adaptable set of principles which acknowledges that ideas must evolve in different contexts. On the other hand, LID's holistic nature, and the existence of core principles, means that to adopt only some of its innovations is to ignore much of what it has to offer and to effectively 'water down' its potential for low carbon living (Maxey and Pickerill 2009). Thus, it is both a 'blueprint' and a patchwork of ideas. This tension is also evident in debates about transition.

We can conceive of transition in multiple ways, but two approaches are particularly relevant here. First, from an economic or management approach, transition can describe processes of social change, with a beginning and an end achieved using a particular strategy – a process with a clear goal. In recent decades, the most common discourses of transition are those proposing 'growth', 'progress' and 'development' (which sit uneasily with recent discussions of 'sustainability transitions'). This is how it has been used in much economic literature, especially when considering transition in post-socialist states (Bradshaw and Stenning 2004).

From a management perspective, using different terminology and more progressively focused on encouraging sustainable transitions, comes the notion of socio-technical transition. This work argues that technological and social change happens when a niche idea is able to intervene in the dominant socio-technical regime (Geels 2005; Geels and Schot 2007). Niches enable innovation, which can occasionally, in complex and uncertain ways, trigger fundamental change (Smith 2007). Using this approach we can identify how certain path dependencies are created (for example, reliance upon oil because of poor public transport infrastructure) but also how opportunities are created for niche ideas to be mainstreamed – such as straw bale building techniques, often used in LIDs (Seyfang 2009; Haxeltine and Seyfang 2009). This approach links the material and the social; it is concerned with the confluence of the technology with the socio-technical practice such that 'social processes shape development and use of technology; but artefacts in turn open up possibilities for new social practices' (Smith and Stirling 2008: 6). Understanding transitions through an examination of how nascent alternatives displace incumbent technologies and practices is important in illuminating the interrelationship between small-scale ideas and systems-level change (Foxon *et al.* 2008).

Thus, we could consider cities as requiring transformation to enact a certain vision, the goal of which can be prescribed in advance. The socio-technical approach adopts quite a bounded notion of place and scale, is often led by elites, and prioritises certain forms of sustainability (Shove and Walker 2007; Walker and Shove 2007). As such, however, it raises pertinent questions about power and agency, about how ideas are translated into 'mainstream' settings, and about accessibility and socially just transitions, which LIDs try to incorporate in their form of social change. Existing work, particularly within geography, has done much to challenge the notion that transition is a neutral or universally beneficial act. For example, work on youth transitions and on post-socialist transitions has explicated the hegemonic political aspirations behind seemingly apolitical programmes for change.

These economic and management approaches require the adoption of certain premises, for example accepting structural causes of our current environmental predicament, in order to clearly identify what needs changing and how. Such approaches tend to ignore local specificities, specific geographies and histories, and the ways in which change is 'constituted through the cultural and social contexts of different time-spaces' (Bradshaw and Stenning 2004: 13). Moreover, they rely on a belief that what existed before – evidence of local cultures, practices and traditions – simply comprised obstructions to a better future. Hence, these economic and management approaches, while being insightful in identifying the importance of innovation, miss much of the complexity, uncertainty and grassroots power that LIDs illustrate in their formation and replication across Britain.

Alternatively, a social movement transitions approach can enable us to do something more complex and more useful: to reconsider how change happens (Begg 2000). It enables us to understand that change is not linear and that we are in a process of constant and dynamic change, of multiple and complex processes of adoption and adaptation. We need to rethink how we conceive of these processes, but they pose an important challenge to understanding transition as only being linear. We can conceive of such transition as a period of interstitiality – the bit 'in between' now and where we would like to be. But it is likely to be forever incomplete and never-ending. Hence, the end point of transition is always unknowable. This sense of unknowable completion critiques the transitions management and economic transitions approach, which suggests that you can manage transitions towards concrete political or economic ends.

Moreover, it leaves open many questions as to who participates and shapes transition as a process and as to the 'location' of transition – between the self (individual) and the social (collective). In these ways, while transition might need to be encompassing and holistic (requiring broad and fundamental changes to our current practices), we can also acknowledge that small changes, which might at first appear apolitical, can change us in unknown ways and that there is 'the potential of social change in unlikely places' (North forthcoming a). This more flexible, less predetermined approach also reduces the need to identify structural causes of societal problems before transition can begin. Although there may well be structural issues at the core of unsustainable practices, a focus on them alone can limit effective collective action. For example, the Trapese Collective (2008) critiqued Transition Towns[4] as failing to acknowledge that the structural cause of climate change is capitalism. As a result, they argue, its focus on local community work is dangerously apolitical and unable to challenge existing power structures. Transition in this sense becomes reliant upon the adoption of an all-encompassing critique of capitalism as a first step towards change. This ignores the value and importance of some of those local structures of voluntary community work and of specific local histories as potential 'building blocks' of the future. Moreover, as much as we might wish not to, we have to begin in the messiness of a capitalist system.

Although a non-structural approach could be apolitical, slow and reformist, to achieve such transition would require major shifts to occur in national political structures and resource provision (Hopkins 2009). Thus, a social movement transition approach acknowledges that we all start from different points and that we do not need to have all the answers before we 'begin'. Moreover, it acknowledges that actually the process of transition – the unknowable, unsettling, unsure space of being in a period of change – can be incredibly productive as a site of creativity, innovation and experimentation (Sargisson 1996: 107; Kraftl 2007) – that it is only through being in that messy period that we might conceive further useful ideas.

Taking a social movements transition approach enables us to look at LID not as an idea that has so far failed to penetrate urban spaces effectively (in other words, failed to penetrate the regime beyond its niche), but as an innovation that pushes for change from the margins. The practical questions LID raises about the possibility of environmentally sustainable cities are important, but, more than that, it enables us to see the futility of a blueprint, the necessity of change on a multitude of levels, and the complexity of engaging in radical social change.

This approach to transition takes much from understandings of how quests for autonomy work in the city. We can consider autonomy as a relational construct where there are no clear boundaries between autonomous and non-autonomous processes and space (DeFilippis 2004). Instead, it involves a constant *negotiation* between competing tendencies towards autonomy and non-autonomy. Through this perspective, transition is not a simply an act of change but part of a complex, negotiated interaction across different places, scales and times. A politics of transition does not involve a linear progression towards an idealised place-bound utopia, but a recognition of the coexistence of different others (who might not accept capitalism as a structural cause), negotiations and conflict. Thus, we could understand spaces of transition as 'entanglements and configurations of multiple trajectories, multiple histories' (Massey 2004: 148), and processes of change that emerge from such spaces as all the more convincing for their complexity.

This second approach to transition is reflected in, and enables a better examination of, LIDs. The seemingly conflicted nature of their role as both 'blueprint' and a patchwork of ideas, of adoption and failure within urban spaces, of unknown starting points, is part of this messy and uneven transitions processes. LIDs are visions of a better future, and a multitude of prefigurative acts that serve both as practical solutions and as inspiring models to be replicated. Importantly, they move beyond simply being environmentally efficient and tackle the broader capitalist constructs at the root of our problems. Such an approach allows us to value a range of different aspects of transition, while not losing site of its urgency and radical possibilities.

LIDs, and their practical manifestation across Britain, are important exemplars of low carbon living. They illustrate the need to embrace radical innovations in this period of climatic (and economic) uncertainty (Seyfang and Smith 2007). Their core principles of the need to adjust our style of housing, our work practices, our personal behaviour, and ensure that these choices are affordable, can be applied in the urban context. Yet there are few examples of urban LID. Despite radical innovative solutions to our many environmental and social problems, tensions remain – political, legislative, economic and personal. However, we can use the example of LID to explore the nature of transition per se. It enables us to understand transition as relational and complex, with unknown starts and unlikely completion, without denying the urgency and necessity for change. There remains much further work to be done in empirically understanding urban LID projects and politically pushing for low carbon projects that are accessible, inclusive and numerous.

Notes

1 I use the term Low Impact Development after Fairlie (1996) to refer to a radical approach to living, as opposed to use of the term in the United States to refer to a specific form of storm-water management system that seeks to disperse storm water using biologically inspired design. Such an approach in Britain is referred to as Sustainable Urban Design Systems (SUDS) and is very different from what is explored in this chapter.

2 Urban spaces are obviously important sites for change, given the concentration of populations, but are broadly considered here as spaces where land is at a premium and high-density housing is required (Pearce 2006).

3 An earthship is a structure with walls made from old car tyres, with waste cans and bottles and earth used to turn them into solid walls. It was developed by Mike Reynolds in New Mexico (Reynolds 1990).

4 Transition Towns (see Chapter 11 of this book), also known as the Transition Movement, are a social movement where communities collectively decide to prepare for the consequences of peak oil and to mitigate the impact of climate change. This tends to involve groups of campaigners raising awareness of environmental issues within their local town, practical efforts to reduce energy use, and attempting to build resilience to climate change by, for example, establishing communal allotments (North forthcoming b). Transition Towns are a complementary campaign, but ultimately separate from LIDs.

References

Anon. (2009) 'Kew Bridge Ecovillage', *Peace News*, no. 2511–2512, July/August.

Begg, A. (2000) *Empowering the Earth: Strategies for Social Change*, Totnes, UK: Green Books.

Bradshaw, M. and Stenning, A. (eds) (2004) *East Central Europe and the Former Soviet Union: The Post-Socialist States*, Harlow, UK: Pearson, Prentice Hall.

BRESCU (2002) *BedZED – Beddington Zero Energy Development*, Sutton General Information Report 89, *Energy and Efficiency Best Practice in Housing*. Online, available at: www.bioregional.com/news-views/publications/bedzedbestpracticereportmar02/.

Brown, G. and Pickerill, J. (2009) 'Space for emotion in the spaces of activism', *Emotion, Space and Society* 2 (1): 24–35.

Chapter 7 (2003) *Sustainable Homes and Livelihoods in the Countryside*, South Petherton, Somerset, UK: Chapter 7.

Corkindale, J. (2007) 'Planning gain or missed opportunity? The Barker review of land use planning', *Economic Affairs* 27: 46–51.
Dale, S. (2009) 'Why do we need Low Impact Development?', in J. Pickerill and L. Maxey (eds) *Low Impact Development: The Future in Our Hands*. Published online at http://lowimpactdevelopment.wordpress.com.
Dawson, J. (2006) 'BedZED and Findhorn, how do they compare?', *Permaculture Magazine* no. 49: 48–52.
DeFilippis, J. (2004) *Unmaking Goliath: Community Control in the Face of Capital Mobility*, London: Routledge.
Fairlie, S. (1996) *Low Impact Development: Planning and People in a Sustainable Countryside*, Chipping Norton, UK: Jon Carpenter.
Fairlie, S. (2003) 'Planning for change: Planning and sustainability have yet to make good bedfellows', *Permaculture Magazine* 36: 17–20.
Featherstone, D. (1997) 'Reimagining the inhuman city: the Pure Genius land occupation', *Soundings: A Journal of Politics and Culture* 7: 45–60.
Foxon, T., Stringer, L. C. and Reed, M. S. (2008) 'Approaches to governing long-term social-ecological change: Comparing adaptive management and transition management', *Ökologisches Wirtschaften* 2: 20–22.
Geels, F. W. (2005) *Technological Transitions and Systems Innovations: A Co-evolutionary and Socio-technical Analysis*, Cheltenham, UK: Edward Elgar.
Geels, F. W. and Schot, J. (2007) 'Typology of sociotechnical transition pathways', *Research Policy* 36: 399–417.
Girardet, H. (2007) *Creating Sustainable Cities*, Schumacher Briefings, Totnes, UK: Green Books.
Halfacree, K. (1999) ' "Anarchy doesn't work unless you think about it": Intellectual interpretation and DIY culture', *Area* 31: 209–220.
Halfacree, K. (2006) 'From dropping out to leading on? British counter-cultural back-to-the-land in a changing rurality', *Progress in Human Geography* 30 (3): 309–336.
Hall, M. and Purchase, D. (2006) 'Building or bodging? Attitudes to sustainability in the UK public sector housing construction development', *Sustainable Development* 14 (3): 205–218.
Hardy, D. (2004) 'BedZED: Eco-footprint in the suburbs', *Town and Country Planning* 73 (1): 29–31.
Haxeltine, A. and Seyfang, G. (2009) *Transitions for the People: Theory and Practice of 'Transition' and 'Resilience' in the UK's Transition Movement*, Tyndall Centre for Climate Change Research, University of East Anglia, Norwich, UK, Working Paper 134, July.
Heeks, A. (2007) 'Cohousing', *Permaculture Magazine* 52: 23–26.
Hewitt, M. and Telfer, K. (2007) *Earthships: Building a Zero Carbon Future for Homes*, Watford, UK: HIS BRE Press.
Holmgren, D. (2002) *Permaculture: Principles and Pathways beyond Sustainability*, Hepburn, Victoria: Holmgren Design Services.
Hopkins, R (2009) 'Transitions Towns: A response', *Peace News*, January: 14.
Kanaley, D. (2002) 'Ecovillages can help solve social and planning issues', in H. Jackson and K. Svensson (eds) *Ecovillage Living: Restoring the Earth and Her People*, Totnes, UK: Green Books.
Kraftl, P. (2007) 'Utopia, performativity and the unhomely', *Environment and Planning D: Society and Space* 25: 120–143.
Laughton, R. (2008) *Surviving and Thriving on the Land: How to Use Your Time and Energy to Run a Successful Smallholding*, Totnes, UK: Green Books.

Lazarus, N. (2003) *Beddington Zero (Fossil) Energy Development: Toolkit for Carbon Neutral Developments – Part II. BioRegional*. Online, available at: www.bioregional.com/news-views/publications/toolkitforcarbonneutraldevelopmentspart2oct03/.

Lovell, H. (2005) 'Supply and demand for low energy housing in the UK: Insights from a science and technology studies approach', *Housing Studies* 20 (5): 815–829.

Low, N., Gleeson, B., Green, R. and Radovic, D. (2005) *The Green City: Sustainable Homes, Sustainable Suburbs*, London: Routledge.

Massey, D. (2004) *For Space*, London: Sage.

Maxey, L. and Pickerill, J. (2009) 'Low Impact Development in the city', in J. Pickerill and L. Maxey (eds) *Low Impact Development*. Published online at http://lowimpactdevelopment.wordpress.com.

Meltzer, G. (2005) *Sustainable Community: Learning from the Cohousing Model*, Victoria, BC: Trafford.

Miles, M. (2008) *Urban Utopias: The Built and Social Architectures of Alternative Settlements*, London: Routledge.

Newton, J. (2008) *Sustainable Home Case Study: Springhill Co-Housing (Stroud, England)*. Online, available at: www.brass.cf.ac.uk/uploads/Sus_Community/Case_Study_Co_Housing.pdf.

North, P. (forthcoming a) 'A rocky critique of transitioning', *Fourth World Review*.

North, P. (forthcoming b) 'The climate change and peak oil movements: Possibilities and obstacles', *Environmental Politics*.

Pearce, F. (2006) 'Ecopolis now: Forget the rural idyll. Urban living may be the best way to save the planet'. *New Scientist* 2556 (17 June): 36–42.

Pickerill, J. and Maxey, L. (eds) (2009a) *Low Impact Development*. Published online at http://lowimpactdevelopment.wordpress.com.

Pickerill, J. and Maxey, L. (2009b) 'Geographies of sustainability: Low Impact Developments and radical spaces of innovation', *Geography Compass* 3 (4): 1515–1539.

Reynolds, M. (1990) *Earthship*, vol. 1, Taos, NM: Solar Survival Press.

Rydin, Y., Amjad, U. and Whitaker, M. (2007) 'Environmentally sustainable construction: Knowledge and learning in London planning departments', *Planning Theory and Practice* 8 (3): 363–380.

Sargisson, L. (1996) *Contemporary Feminist Utopianism*, London: Routledge.

Schwarz, W. and Schwarz, D. (1998) *Living Lightly: Travels in Post-consumer Society*, Chipping Norton: Jon Carpenter.

Seyfang, G. (2009) *The New Economics of Sustainable Consumption: Seeds of Change*, Basingstoke, UK: Palgrave Macmillan.

Seyfang, G. and Smith, A. (2007) 'Grassroots innovations for sustainable development: Towards a new research and policy agenda', *Environmental Politics* 16 (4): 584–603.

Shove, E. and Walker, G. (2007) 'CAUTION! Transitions ahead: Politics, practice and sustainable transitions management', *Environment and Planning A* 39: 763–770.

Smith, A. (2007) 'Translating sustainabilities between green niches and socio-technical regimes', *Technology Analysis and Strategic Management* 19 (4): 427–450.

Smith, A. and Stirling, A. (2008) 'Social-ecological resilience and socio-technical transitions: Critical issues for sustainability governance', STEPS Working Paper 8, Brighton: STEPS Centre.

Trapese Collective (2008) *The Rocky Road to a Real Transition: The Transition Towns Movement and What It Means for Social Change*. Online, available at: http://sparror.cubecinema.com/stuffit/trapese/.

Twinn, C. (2003) 'BedZED', *Arup Journal* 1: 10–15.

UWE (University of the West of England)/Land Use Consultants (2002) *Countryside Council for Wales. Low Impact Development – Planning Policy and Practice. Final Report*. Bristol: UWE/Land Use Consultants.

Walker, G. and Shove, E. (2007) 'Ambivalence, sustainability and the governance of socio-technical transitions', *Journal of Environmental Policy & Planning* 9 (3–4): 213–225.

Ward, C. (2002) *Cotters and Squatters: Housing's Hidden History*, Nottingham: Five Leaves.

Wolch, J. (2007) 'Green urban worlds', *Annals of the Association of American Geographers* 97 (2): 373–384.

Wrench, T. (2001) *Building a Low-Impact Roundhouse*, East Meon, Hampshire, UK: Permanent Publications.

13 Conclusion

Mike Hodson, Simon Marvin, Harriet Bulkeley and Vanesa Castán Broto

The chapters of this book constitute a systematic and far-reaching attempt to understand low carbon urban transitions. The book has brought together contributions from different disciplinary, geographical, political and epistemological positions to illuminate, interrogate and reflect upon a variety of aspects of what the contributors have both explicitly labelled and implicitly referred to as urban transitions. The theoretical, conceptual, methodological and empirical breadth of these contributions allows us to return to the question that we outlined at the beginning of the book: how, why, and with what implications, are cities effecting low carbon transitions? Reflecting on the contributions in relation to this question leads to five critical conclusions that synthesise arguments, critically engage with key issues and open up a future urban transitions research agenda.

The contested politics of urban transitions

Low carbon urban transitions are about competing views of the role of the city, the type of transition that is required, how to translate that vision, and, therefore, the variability of the consequences of a transition. What is clear from the contributions to this book is that urban transitions or transition activity are often dominated by narrow coalitions of actors. Furthermore, low carbon transitions and transitions more generally are often seen to have a degree of inevitability – both in the broader sense of something needing to be done about climate change and also more specifically when that challenge is taken up, for example, by urban political elites or community groups. But the simplicity of this assumption has been exposed by the contributions to this book, with the result that the very practical implications of what a socio-technical transition means – in terms of the breadth of participation and the organisational cultures that are required – have begun to be opened up. The creation of more inclusive forms of transition is challenging in a context where, in many parts of the urban world, extremes of wealth and poverty remain and have been exacerbated and where cities, communities, groups, individuals, utilities, business interests and so on within transitions are very differentially positioned in their capability to act.

The key finding from this book is that transitions are often the subject of narrow social interests and politically contested. This means that there is a requirement for the multilevel perspective (MLP) and transitions research to undertake more critically reflective analysis of the processes and politics of how transitions have been undertaken or will take place. This requires, on the one hand, confronting the seeming inevitability in some historical accounts of transitions, and, on the other hand, extending the often limited practical actions frequently seen as necessary to achieve a future transition to include a broader range of social, technical, economic and political interventions. The critical message of the contributions to the book is that locating the city in low carbon transitions is the subject of politics and struggle. The act of representing a mutual low carbon future for the city and urban infrastructure is underpinned by political efforts to bound time-space through socio-technical transition in the service of particular social interests. In this sense, the MLP provides us with a foil through which not only to think through the role of the city in low carbon transitions but also to understand historically the roles of cities in socio-technical transition. Yet it also requires critical engagement between the MLP and a further set of issues that are explored in the rest of this concluding chapter.

The roles of cities and the multi-scalar production of urban transitions

The contributions to the book collectively highlight that locating the role of the city – theoretically, conceptually and empirically – in low carbon transitions is extremely difficult. The multilevel perspective says little explicitly about cities, who, what and where the city is, and cities' roles within transitions. The chapters detailed some of the many different ways of thinking about the roles of cities in low carbon transitions. This included viewing the city as a transition actor, as a contributor to national-level transitions (Geels), through the lens of decision-making calculus (While), as constituted through multilevel governance coalitions of interest (Späth and Rohracher), the organisational cultures of urban transition (Aylett), the meso-level internal dynamics of urban energy usage (Dhakal), but also through 'alternative' spaces within the city through which often marginalised voices seek to participate in low carbon transitions (Pickerill; Smith). If we look at these different ways of thinking about the role of the city in low carbon transitions, a degree of illumination can be achieved through engagement with and questioning of the MLP – questions that the contributors to this book have posed. To take a few examples: do 'alternative' spaces constitute niches? Can cities solely contribute to national-level transitions or can they constitute regimes at the scale of the city? Can new decision-making calculus contribute to reshaping the regime at a city level or does this inform experimental niche activity?

What is clear from the contributions to this book and also the literatures that they engage with is that there are potentially a multiplicity of ways not only of conceptualising the role of the city in transitions but also of thinking about the relationships of scales that constitute the city to the wider notion of an urban low carbon transition. Whether the city is viewed as an actor, a niche, a regime or any other conceptualisation has important consequences for how a series of contexts and relationships 'within' and 'outside' the city are understood. How communities and publics relate to the city and also to national states is an important issue. To push this further, for example, we could asses whether 'alternative' low carbon transitions are genuinely alternative or whether they have the unintended consequence of legitimising policy-led transitions through not directly questioning and engaging them. To take another example, we could critically evaluate whether and how the disciplining of city authorities, by national government, through imposing targets subsequently feeds into a disciplining of individuals, homeowners, citizens and consumers.

Intermediaries, experimentation and the organisation of transitions

What this diversity of transition initiatives and activities suggest is that a wide range of social interests are staking a claim to speak for the city in undertaking or aspiring to undertake low carbon transitions. The field of such social interests is not an equitable one, and highly particular coalitions of social interests, in relation to particular places, are often able to mobilise financial, relational and knowledge resources through which they are often able to produce the symbolic 'visions' of what the low carbon future of the city should be (Späth and Rohracher). Yet in among this politics that privileges certain interests over others, it is clear that there are potentially many ways of organising transitions. How this can occur relates to the issue of whether such processes can be managed or they simply unfold in messy ways (Bulkeley *et al.*). Underpinning this are different organisational contexts and cultures that mediate and through which social interests and a range of resources coalesce (Hodson and Marvin). Yet this is not a fixed situation, and these organisational cultures not only are likely to differ between transition contexts but also are likely to change over time – and be subject to purposive transformation – in the same transition context (Aylett).

Crucial to understanding these processes are the intermediary organisations and the experimental processes through which these relationships are developed, often in living laboratories or other similar experimental settings, and the politics through which knowledge and action are produced (Evans and Karvonen). It is also critical to understand how these various experimental processes, relationships and the forms of knowledge that they produce relate to existing regimes and whether they support the buttressing and defence of existing regimes or whether they create the possibilities for genuine low carbon transition and what the balance is between transformation and continuities in the organisation of networks (Coutard and Rutherford).

What appears to be the case is that there is often a disjuncture in participative terms – a disjuncture between policy-led responses that are 'externally' facing, in that they take little account of, or fail to engage with, publics in seeking to make urban contexts amenable to inward investment, competition and business, and 'alternatives' that are often guilty of the same privileging in reverse, understandably prioritising local social interests but ignoring the availability of resources and forms of knowledge and technology that may be drawn on from policy contexts and beyond. Broadening our understanding of the social interests and forms of knowledge involved in transitions creates the potential to think about how we would know whether a transition had been achieved through its social embeddedness and acceptability rather than purely focusing on attempts by narrow groups of social interests to design, implement and, in some cases, impose a transition.

Understanding the scope and consequences of transitions activity

What the chapters also illuminate is the tension between whether urban low carbon transitions are manageable – indeed, more manageable than national transitions, given their proximity between actors, networks and place – and can be shaped, or are messy and wrought by the unintentional co-incidence of a multiplicity of actions. This, as the contributions have shown, should not be treated as a zero-sum question. There are significant positions in between, including the view that transitions can be the consequence of purposive intent, unintended consequences and subsequent adjustments. This, then, is to lay down three framings, of potentially many, as to whether urban low carbon transitions can be directed. It also acknowledges the teleological leanings of accounts that view the management of transitions in time-bounded ways rather than viewing time horizons as looser reference points and as part of ongoing and unfolding processes of transition. Yet the framing of low carbon transitions can be manifold, and often, as illustrated by the chapters, the basis for different framings is through their participative constitution by transition actors. To put it another way: Who are involved in transitions and on what basis do they shape the way in which a transition is scoped out and represented to others? The more tangible implications of this can be seen where, for example, low carbon transitions are regarded solely in the realm of urban policy regimes and coalitions or, conversely, where transitions are understood solely at the level of everyday practices. Unpicking the politics of those processes allows us to consider whether there are ways of bringing together purposive policy-led transitions with the variety of contexts of everyday practice. Furthermore, the MLP has primarily been concerned with national-level transitions. There are critical issues raised by the ways in which national governments view the role of cities and communities in undertaking transitions and whether they view such roles as contributing to a national transition or whether a significant degree of devolution to these contexts takes place with the inevitable loss of national control that goes with such processes. The framing of urban low carbon transitions needs a more sophisticated understanding not only of the national state but also of wider political economy and geopolitical pressures and the ways in which these factors combine to create – but not determine – conditions for transition activity in relation to particular cities and communities.

Rethinking transitions

The contributions to this book have provided an engagement with urban low carbon transitions that has enabled us to synthesise the critical issues set out in the previous four sections. As academics who seek not solely to observe the world but also to influence its future, we recognise that further research in this field needs to engage with these issues. More specifically, research engagement needs to recognise the differences encompassed by the term 'urban low carbon transition' – between material contexts, social interests, available resources, relationships to national government and so on. More historical accounts of the changing roles of cities in transitions are required. This necessitates the development of a critical engagement with different transition contexts. This requires more engagement theoretically, conceptually and empirically with whether transitions can be managed as well as the bringing together of historical and future-oriented accounts of urban transition. There is a need through research to work on connecting transition contexts, such as the overtly policy and technical visions of some urban low carbon transitions with more overtly cultural and contextually embedded attempts at transition. Theoretically and conceptually, this needs, and subsequently feeds into, more work on the roles of cities in the MLP beyond the limited characterisations and roles of cities that have been developed hitherto. Methodologically, this means multiple roles for researchers from scholarly historical, political, geographical and sociological accounts of transitions, but also a more practical engagement with unfolding urban low carbon transitions through action research. Addressing such issues promises the beginnings of an ongoing agenda for the low carbon future of cities and their dependent relationships with the very socio-technical networks that sustain them.

Index